TRIGONOMETRY

The Easy Way

THIRD EDITION

Douglas Downing, Ph.D.
Seattle Pacific University

BARRON'S

Dedication

This book is for Bill Trimble

Acknowledgments

I would like to thank my teacher Clint Charlson, my mother Peggy Downing and my father Robert Downing for their help. Dr. Jeffrey Clark and Mark Yoshimi provided inspiration and Marlys Downing provided creative assistance. Mickey Wagner and Linda Turner at Barron's worked hard to edit the book. Special thanks also to Susan Detrich for the marvelous illustrations.

© Copyright 2001 by Barron's Educational Series, Inc.
Prior © copyright 1990, 1984 by Barron's Educational Series, Inc.

All inquiries should be addressed to:
Barron's Educational Series, Inc.
250 Wireless Boulevard
Hauppauge, New York 11788
http://www.barronseduc.com

Library of Congress Catalog Card No. 00-062157

ISBN-13: 978-0-7641-1360-4
ISBN-10: 0-7641-1360-7

Library of Congress Cataloging-in-Publication Data
Downing, Douglas.
    Trigonometry : the easy way / Douglas Downing.—3rd ed.
        p.   cm.
    Includes index.
    ISBN 0-7641-1360-7
    1. Trigonometry. I. Title.
QA531 .D683 2001
516.24—dc21
                                        00-062157

PRINTED IN THE UNITED STATES OF AMERICA
15 14 13 12 11 10

Illustrations by Susan Detrich

Contents

the Professor

Trigonometris

Marcus Recordis

Gerard Macinius Builder

the King

MAIN CONFERENCE ROOM

Introduction

This book tells of adventures that occurred in a faraway fantasy kingdom called Carmorra. During the course of these adventures, the people developed a brand-new subject, *trigonometry*. By reading this book you can learn trigonometry. The book covers material that is studied in a high school or first year college trigonometry course.

Trigonometry started as the study of triangles. Many applications of trigonometry involve solving triangles. Engineers, astronomers, navigators, and physicists all need to know trigonometry. However, trigonometry can also be used to solve many problems that are unrelated to triangles. Oscillating motion, electric current, sound waves, and light waves can all be described by trigonometric functions. You will also find a knowledge of trigonometry essential if you study advanced mathematics, beginning with calculus.

To appreciate this book you should have a bit of knowledge of algebra and geometry. You should be familiar with function notation, such as $y = f(x)$, because we spend most our time in trigonometry studying a special kind of function. You also should know how to identify points with an xy coordinate system. You should have studied enough geometry to be familiar with the degree system for measuring the size of angles, and you should know some of the basic properties of triangles. This material is reviewed in Chapter 1. If you know geometry well, you may wish to skip over Chapter 1 to the beginning of the trigonometry material in Chapter 2.

There are exercises at the end of each chapter to give you practice with the material. Understanding any mathematical material requires work. The answers to the exercises are included at the back of the book so you can check your work. There is no way to avoid some memorization. You should memorize the definitions of the sine, cosine, and tangent functions, and you should memorize the special values for these functions. You should not try to memorize all the important formulas, but you may look these up in the special section at the back of the book.

Stars (★) mark exercises that are more difficult or that require more background knowledge.

In the old days trigonometry was a very tedious subject because of the complexity of the calculations involved. In order to solve a trigonometry problem, you needed to look up the values of the trigonometric functions in a bulky table. These days the work is much easier because you can obtain a calculator that will calculate the values of trigonometric functions at the touch of a button. Most of the exercises in the book are designed to be done with calculators. You should learn how to enter a number into your calculator and then how to calculate the sine, cosine, or tangent of that number.

The computer programming exercises, marked with a box, have been included to illustrate some of the ways in which computers can be used to help solve trigonometry problems. Many problems, such as drawing graphs, are very difficult without the aid of computers. Sample solution programs have been included at the back of the book. These programs are written in BASIC, and they can be adapted for other programming languages.

A computer spreadsheet can also help you solve trigonometry problems. You can enter a formula into a spreadsheet that automatically recalculates when you change the numbers in a problem. You can also use a spreadsheet to create graphs. See the appendix for information on examples in which spreadsheets are used to solve trigonometry problems.

Radian measure for angles is developed in Chapter 5. After that both radian measure and degree measure are used. You should become familiar with both measuring systems, and you should be able to convert from one to the other. In general you will find that radian measure is more convenient for mathematical purposes but degree measure is more convenient for practical purposes when you are measuring angles.

There are a few Greek letters you will need to become familiar with. You should already recognize the Greek letter *pi* (π) as the symbol for the circumference of a circle that has diameter 1 (and you should know $\pi = 3.1416\ldots$). The Greek letter *theta* (θ) is often used to represent angles, but to avoid introducing too many new symbols, we do not use θ until Chapter 11. The other Greek letters we use are *omega* (ω) for angular frequency, *lambda* (λ) for wavelength, and *phi* (ϕ) for angles.

The last half of Chapter 4 and all of Chapter 9 cover applications of trigonometry to physics and music. Chapters 12 to 15 cover material that requires a deeper understanding of algebra topics, such as complex numbers, polynomials, and conic sections. You may omit these chapters if you like. The advanced sections are marked with stars ★.

When you first study trigonometry you are likely to find the subject baffling because of the new symbols used. At the beginning of the story the characters are in the same position you are now. They don't know trigonometry either. During the course of the book, they learn trigonometry, just as you will. Once you become familiar with the trigonometric functions you will see that they make it possible to discover concise, elegant solutions for many problems (although you probably will not come to regard the trigonometric functions with the same degree of devotion shown by Alexanderman Trigonometeris).

This book no longer contains a table of the trigonometric functions because these days you will be using a calculator or computer whenever you need to look up these values. However, if you do need to consult a table to verify that a value has been figured correctly, check the web page for this book at *http://www.spu.edu/~ddowning/easytrig.html*. This web page also has some additional examples of how to use a computer to graph trigonometry problems.

Good luck. You're now about to set out on the journey of learning trigonometry.

1 Angles and Triangles

It rained for days. Everybody in the entire kingdom of Carmorra was forced to stay inside to avoid the drenching downpour. I took refuge in the king's palace along with the other members of the Royal Court. (I had been residing in the palace since I had been stranded in the strange faraway land of Carmorra by a shipwreck.)

Marcus Recordis, the Royal Keeper of the Records, stared dolefully out the Main Conference Room window. "Rain rain, go away; come again some other day," he sighed. He watched the rainwater slide off the sloped roof of the palace. "I think we should change the tilt of the roof," he remarked. "If we made the roof steeper, then the water would run off the roof more easily."

Figure 1-1

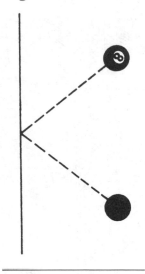

**Measuring
Angles**

Gerard Macinius Builder, the Royal Construction Engineer, looked up from the drawings he was using to plan his latest building project. "If you want to change the steepness of the roof, you will have to be very specific and tell me precisely how much tilt you want."

"I don't known how to measure the amount of tilt of a roof," Recordis complained. But Builder had other problems. He was staring at his drawings in puzzlement.

"Most of the walls in this building will meet to form square corners," Builder said. "However, at one location two walls will meet but they will not form a square corner. Before I can proceed I must have a way to precisely measure the angle between two walls."

Professor Stanislavsky, the country's leading pure scientist, was relaxing by practicing pool. She prepared to take aim for a particularly tricky shot. (See Figure 1-1.)

"I need to hit the cue ball against the wall and have it bounce back to hit the eight ball," she explained to Recordis. "I need to calculate my direction of aim. I wish we had a more precise way to measure directions."

The King of Carmorra had been staring thoughtfully out the window listening to the conversation. His face was careworn from the pressures of being a fair ruler. He had led the kingdom through many exciting moments. Some of the most memorable adventures had occurred while we were discovering the subject of algebra. Finally, an idea struck him. "I know how to find the solution to all these problems," the king announced. "We need a way to measure angles!"

"First we had better define precisely what we mean by the word *angle*," Recordis said. He pulled out one of his trusty notebooks. Since his job required him to keep a written record of every significant event that happened at the royal court, he always kept several notebooks at his side and several pens and pencils stuck behind his ear.

"That's easy," the professor said. "An angle is a place where two lines cross each other." She drew a picture. (See Figure 1-2.)

"It looks to me as if a crossing place between two lines forms four angles," Recordis said.

"To avoid that problem we will say that an angle is a place where the end points of two *rays* meet each other," the king said. "Remember that a ray is like half a line. A line goes off to infinity in two directions, but a ray has one ending point and then it goes off to infinity in one direction." (See Figure 1-3.)

Figure 1-2

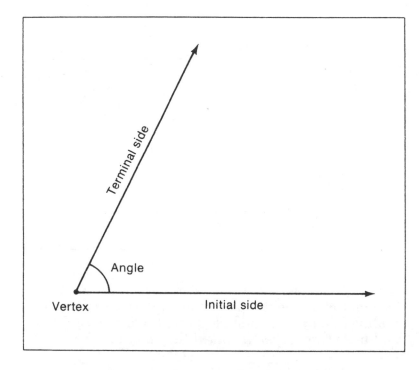

Figure 1-3

"The beam of light from Pal's toy ray gun is like a ray," Recordis remembered. "The beam starts at the gun and then goes off to infinity in a straight line." (Pal was a friendly giant who often helped the people of Carmorra when they were in trouble.)

"We'll call the point where the two rays meet the *vertex* of the angle," the professor suggested. She liked to make up new names for new things. "We will call one of the rays the *initial side* and the other ray the *terminal side.*"

We drew some angles. (See Figure 1-4.)

"Some angles are very sharp and other angles are very blunt," Recordis said.

Figure 1-4

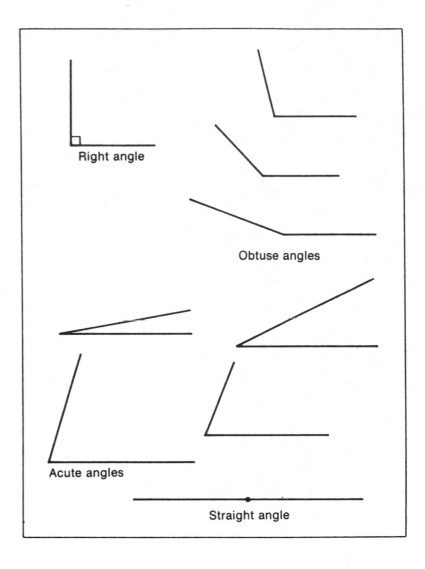

Right angle

Obtuse angles

Acute angles

Straight angle

"We should call a square corner a *right angle*, since that is the right type of angle to use when you are building a house," Builder suggested. We also said that two lines were *perpendicular* if they met to form a right angle.

We decided to use the term *acute angle* for an angle sharper than a right angle. We also coined the term *obtuse angle* for an angle larger than a right angle.

"You get a very strange angle if the two rays point in opposite directions," the professor said. "In that case the angle looks as if it is really a straight line, so we should call it a *straight angle*."

"We can measure angles by stating what fraction of a straight angle the angle fills," the king suggested. "Then a straight angle would measure 1, a right angle would measure $\frac{1}{2}$, and so on."

"That method will involve too many fractions!" Recordis complained. "I would much prefer a system in

which the most commonly used angles, such as half
of a straight angle, one-third of a straight angle, and so
on, are all represented by whole numbers. Let's pick a
big number that is divisible by lots of other numbers to
represent a straight angle." Recordis decided that he
wanted to use a number divisible by all these numbers:
2, 3, 4, 5, 6, 9, 10, 12, and 15. After some calculation
we found that 180 was the smallest number divisible
by all these numbers, so the king issued a Royal
Decree.

> A straight angle will have a measure of
> 180 degrees, which we will write as 180°. A
> right angle measures 90°; an angle that is one-
> quarter of a straight angle measures 45°; an
> angle that is one-sixth of a straight angle
> measures 30°; and so on. (See Figure 1-5.)

(The professor had pointed out that we needed a
name for the units we were using to measure angles, so

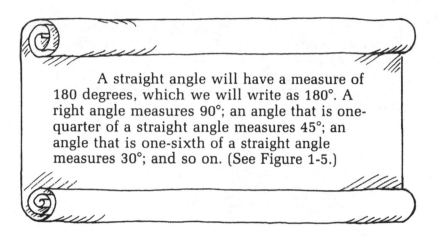

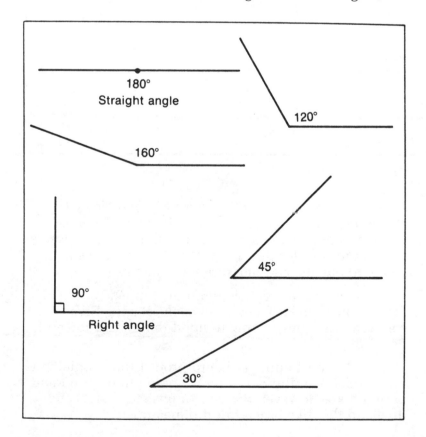

Figure 1-5

Recordis suggested the name *degree* because he liked detective novels in which people were charged with counts to the first degree and counts to the second degree. We decided to use a little raised circle ° as the symbol to represent degrees. We later found that it was useful to develop another system for measuring angles, called *radian measure*. According to radian measure, a straight angle measures π, where π is a symbol for a special number that is about equal to 3.14159. See Chapter 5. In radian measure a right angle measures $\pi/2$.)

"Now we need to invent a device that we can use to measure angles," the professor said. After some discussion, we designed a device shaped like a semicircle with numbers running from 0 to 180 along the edge. We called this device a *protractor*. (See Figure 1-6.)

Figure 1-6

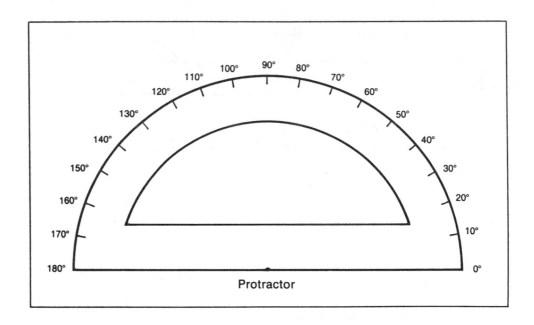

Protractor

The professor wanted to try her new device, so she quickly measured every angle in sight. Soon she had measured every single angle in the Royal Palace, so she had to wait for the end of the storm before she could go outside and measure more angles.

However, as soon as the rain stopped, catastrophe struck. The waters of the Raging River started to rise, threatening to flood the distant town of Peaceful Bay.

The Raging River Flood

"We must build a dike to protect the people!" the king exclaimed. Builder quickly sprang into action and built a dike. However, the waters pushed against the dike and the dike began to tilt dangerously.

"I need a more rigid shape!" Builder cried. "What is the most rigid shape in the world?"

We quickly constructed several test shapes. We constructed squares, pentagons, hexagons, decagons, and many others. Each shape had unbending sides but flexible hinges at each vertex, and Pal had no trouble bending each one out of shape.

"There's only one shape we haven't tried yet," Recordis said breathlessly. "We haven't tried the simplest shape of all—a triangle."

"A triangle?" the professor exclaimed skeptically. However, Builder quickly constructed a triangle and Pal was unable to bend it out of shape.

Triangles to the Rescue

"The triangle is perfectly rigid!" Builder said in astonishment. "This is a fundamental fact that will help with the design of many different construction projects."

Builder quickly constructed a triangularly shaped support tower for the dike, and the waters were brought under control. The town was saved.

In honor of the triangle, we constructed a new park in the middle of Capital City called Central Plaza Triangle. The professor suddenly became interested in the entire subject of triangles. Previously she had scorned triangles as being too simple to be worthy of serious scientific investigation. However, during the next week she conducted a very detailed investigation of all types of triangles. "There are more different types of triangles than you might imagine," she said. She decided to write a book on triangles. The writing process took a long time, since she told us that she spent hours contemplating each word. Finally she was finished, and we were all impressed when she gave us the first copy of the book to put in the Royal Library. She graciously gave me permission to reprint the book here.

Complete Guide to Everything Worth Knowing About Triangles

by Professor A. A. A. Stanislavsky, Ph.D., etc., etc.

A triangle consists of three line segments joined together end to end. The three points where the line segments meet are called the vertices. The three line segments are called the three sides of the triangle. A triangle contains three angles.

Complete Guide to Triangles

If you add together the three angles in any triangle, the result will be 180°.

The area of a triangle is equal to $\frac{1}{2} \times$ base $\times$ altitude. You may call one of the three sides the base. Then the altitude is the perpendicular distance from the base to the opposite vertex.

If the three sides of a triangle are equal, then it is called an *equilateral triangle*. An equilateral triangle contains three 60° angles.

If two sides of a triangle are equal, then it is called an *isosceles triangle*. In an isosceles triangle, the two angles opposite the two equal sides will be equal to each other.

If the three sides of a triangle are all unequal, then it is called a *scalene triangle*.

If a triangle contains one 90° angle, then it is called a *right triangle*. The longest side of a right triangle is called the *hypotenuse*. It is the side opposite the right angle. The two other sides are called the *legs*. The two other angles must add up to 90°. (If two angles add up to 90°, then they are said to be *complementary angles*.)

If c is the length of the hypotenuse, and a and b are the lengths of the two legs, then

$$c^2 = a^2 + b^2$$

(This result is known as the *Pythagorean theorem*.)

If all three angles of the triangle are less than 90°, then it is called an *acute triangle*. If one angle is greater than 90°, then it is called an *obtuse triangle*. (See Figure 1-7.)

(A triangle cannot have more than one obtuse angle because the sum of the angles in a triangle must always equal 180°.)

Two triangles are *congruent* if they have the same shape and size. If you could pick up one of the triangles and put it on top of the other, then the two triangles would fit together perfectly. Let's call one of the triangles "triangle 1" and the other triangle "triangle 2." Each side of triangle 1 is the same length as its corresponding side on triangle 2. Each angle of triangle 1 is the same size as its corresponding angle on triangle 2.

Two triangles with the same shape but different sizes are said to be *similar triangles*. For example, the real Central Plaza Triangle is exactly the same shape as the picture of Central Plaza Triangle on Recordis's map of Capital City.

Figure 1-7

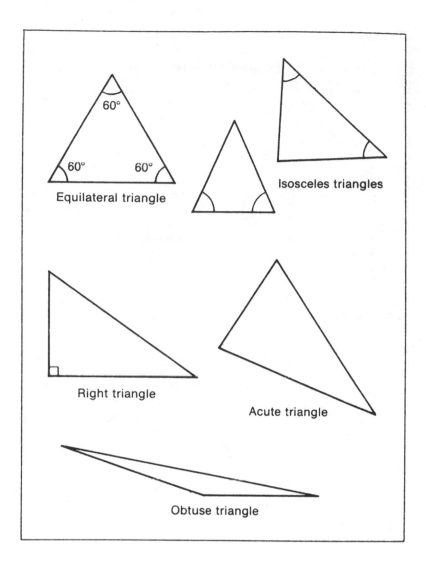

Equilateral triangle

Isosceles triangles

Right triangle

Acute triangle

Obtuse triangle

However, the real triangle is obviously much larger than the triangle on the map. (See Figure 1-8.)

Let's imagine that we are looking at any pair of similar triangles. Each angle on the big triangle is the same size as its corresponding angle on the little triangle. Now, let's compare the length of each side of the big triangle with the length of its corresponding side of the little triangle. Let's imagine that one side of the big triangle is 10 times longer than its corresponding side on the little triangle. That means that *all* the sides of the big triangle will be 10 times longer than their corresponding sides on the little triangle. Or, if one side is twice as long as its corresponding side, then all the sides will be twice as long as their corresponding sides. This means that the corresponding sides of similar triangles have the same proportion.

Figure 1-8

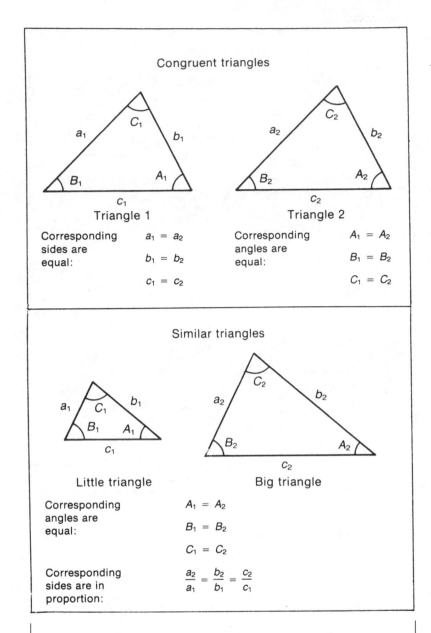

Congruent triangles

Triangle 1

Triangle 2

Corresponding sides are equal:

$a_1 = a_2$

$b_1 = b_2$

$c_1 = c_2$

Corresponding angles are equal:

$A_1 = A_2$

$B_1 = B_2$

$C_1 = C_2$

Similar triangles

Little triangle

Big triangle

Corresponding angles are equal:

$A_1 = A_2$

$B_1 = B_2$

$C_1 = C_2$

Corresponding sides are in proportion:

$\dfrac{a_2}{a_1} = \dfrac{b_2}{b_1} = \dfrac{c_2}{c_1}$

(We used capital letters for angles and lower case letters to represent the lengths of the sides.)

Triangles are very useful for construction purposes because they are rigid. That means that a triangle with rigid sides but flexible hinges cannot be bent out of shape. Any other polygon with rigid sides but flexible hinges can be bent out of shape.

Little did we suspect at the time that this was just the beginning of a rather remarkable set of adventures. We started out studying triangles, but along the way we made many other discoveries that were only slightly related to triangles.

- We gave a special name to an angle with a vertex at the center of a circle. This type of angle is called a *central angle*. (See Figure 1-9.) Note that the two sides cut across the circle. The piece of the circle between the two sides is called an *arc*.

Figure 1-9

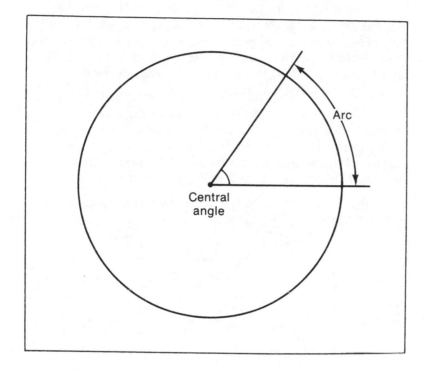

- Suppose we need to measure a very small angle. For example, we might want to measure an angle that measures 0.05°. In that case we can express the angle in terms of *minutes*, where 1 minute $= \frac{1}{60}$ of a degree. Therefore,

$$1 \text{ minute} = 0.0167°$$

and

$$3 \text{ minutes} = 0.05°$$

If we need to measure very, very small angles, then we can express the angle in terms of *seconds*, where 1 second $= \frac{1}{60}$ of a minute. Therefore,

$$1 \text{ second} = \frac{1}{3600} \text{ degree} = 0.0002778°$$

(Don't confuse the minutes and seconds discussed here with the minutes and seconds we use to measure time. It is unfortunate that the same terms are used for both. Sometimes the phrase "minutes of arc" or "seconds of arc" is used when we are referring to angular measure.)

When writing a very small number, it is convenient to use *scientific notation*. In scientific notation, the number 0.0002778 is written as 2.778×10^{-4}. A number in scientific notation is expressed as the product of a power of 10 (in this case, 10^{-4}) multiplied by a number between 1 and 10 (in this case, 2.778.)

Scientific notation is also used for big numbers. For example, 2,340,000,000 can be written as 2.34×10^9. When using computers, this notation is often called *exponential notation*, and it is written as 2.34E9. (The E stands for exponent; the number following the E indicates the power of 10.)

Take a moment to learn exactly how your calculator displays exponential notation. The display might look like 2.34e9, or it might be $2.34^{\,9}$, or the exponent might appear off to the side next to an indicator letting you know it is the exponent.

Exercises

For Exercises 1 to 8, fill in the missing elements in the table for right triangles.

	Short leg	Long leg	Hypotenuse
1.	3	—	5
2.	6	8	—
3.	—	12	13
4.	7	24	—
5.	1	1	—
6.	1	3	—
7.	41.955	—	65.27
8.	—	2.9544	3

Find the angle that is complementary to each of the angles in Exercises 9 to 14.

9. 45°

10. 30°

11. 60°

12. 75°

13. 90°

14. 22.5°

For each of the triangles in Exercises 15 to 20, two angles are given. Calculate the size of the third angle.

15. 45°, 45°

16. 30°, 90°

17. 60°, 60°

18. 10°, 10°

19. 100°, 70°

20. 20°, 90°

The angle between the two equal sides of an isosceles triangle is given for Exercises 21 to 25. Calculate the size of the other two angles.

21. 100°

22. 80°

23. 90°

24. 140°

25. 40°

★26. What will be the sum of the angles in a quadrilateral?

★27. What will be the sum of the angles in a pentagon?

★28. What will be the sum of the angles in an n-sided polygon?

★29. Prove that the sum of the angles in a triangle is 180°.

★30. Prove that, in an isosceles triangle, the two sides opposite the two equal angles are equal to each other.

Convert the angles in Exercises 31 to 34 from degrees to degrees-seconds-minutes.

31. 16.5°

32. 22.333°

33. 2.22×10^{-3} degrees

34. 0.202°

Convert the angles in Exercises 35 to 38 from degrees-seconds-minutes to decimal degrees.

35. 12° 15 minutes

36. 34° 50 minutes

37. 4 seconds

38. 5° 14 minutes 4.8 seconds

2
Solving Right Triangle Problems

The Height of the Tree

Long before Christmas the members of the royal court began making plans for the large Christmas tree to be displayed in Central Plaza Triangle. We went into the forest and found just the tree we wanted. Builder, as usual, prepared to do the actual work involved with cutting down the tree and setting it up. However, he needed to know the height of the tree before he could begin. (Figure 2-1 illustrates the situation.)

"We're in real trouble now!" Recordis exclaimed. "There is no way that I can climb that tree with my tape measure! I can easily measure flat things, but not trees!" To prove that he still could measure some things, Recordis stretched out his tape measure and determined that the shadow of the tree was exactly 50 feet long. "However, I don't see how this information is going to help us," he said glumly.

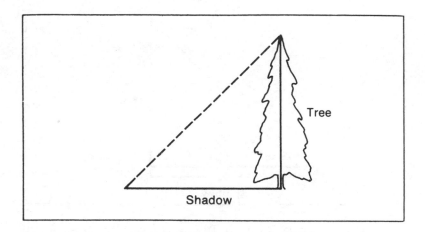

Figure 2-1

"Do we have any other information?" the professor asked. "I always say, 'When confronted by a difficult problem, the more information you have, the better.'"

The king paced nervously back and forth. He did not want to be the one to tell the townspeople that there would be no tree this year. Recordis didn't have anything else to do, so he decided to measure the length of the king's shadow. "This is interesting," he said. "Your shadow is exactly as long as you are tall. This means that the triangle formed by you, your shadow, and the line connecting the tip of your shadow to the top of your head is an isosceles triangle."

"That means that the sun's angle of elevation is exactly 45°," the king said.

"Therefore, the angle formed by the ground and the line joining the tip of the shadow to the top of the tree must measure 45°," the professor observed. "That might be an important clue." (See Figure 2-2.)

"This situation looks familiar," Recordis said.

"I know!" the king said. "We know that the tree

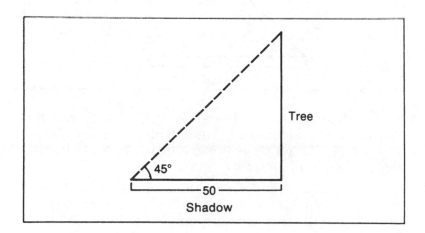

Figure 2-2

Figure 2-3

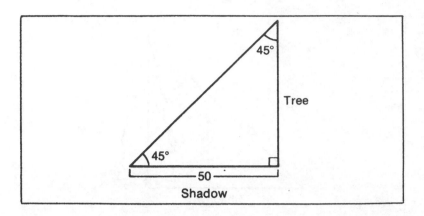

trunk forms a right angle with the ground. Therefore, the triangle formed by the tree, the ground, and the line connecting the top of the tree to the tip of the shadow is a right triangle." (See Figure 2-3.)

We marked a small square on the diagram to show the location of the right angle.

"Since the sum of the angles of any triangle add up to 180°, we know the other angle in the triangle must also be 45°," the professor said helpfully.

Recordis added, "We know that two sides of an isosceles triangle are equal. (I dare you to draw a triangle that has two equal angles but with the sides opposite those angles not equal.)" His eyes widened as he suddenly realized the implications of what he had just said. "Therefore, the height of the tree must be equal to the length of the shadow—in other words, the tree must be 50 feet high!" (See Figure 2-4.)

Figure 2-4

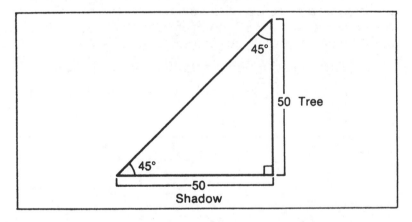

"The problem has been solved!" the king said gladly. "Now every town in the kingdom will be able to have a tree. We can write down a general procedure to find the height of any tree."

1. Walk away from the tree until you reach the point where the angle of elevation of the top of the tree as seen from the ground is 45°.

2. Measure the distance from the tree to that point.

3. The height of the tree is equal to that distance.

"Let's state this method in a bit more general terms," the professor said. While she was studying algebra she had learned the value of writing the solution to a problem as generally as possible. Then the same solution method could often be used for many different problems, thereby saving a lot of work. "In any particular right triangle, we may choose one of the nonright angles, which we will call the *angle of interest.* As we have seen, the longest side of the right triangle is called the hypotenuse. It is opposite the right angle. The short side that touches the angle of interest will be called the *near side* or the *adjacent side.* The other side of the triangle will be called the *far side* or the *opposite side.* (See Figure 2-5.)

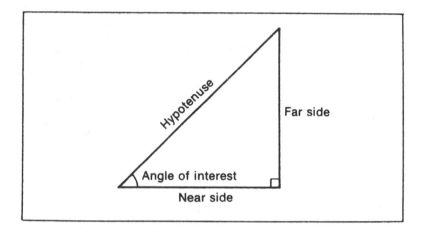

Figure 2-5

"Then we may state one general result that will help us with triangle problems."

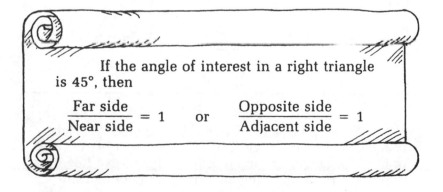

If the angle of interest in a right triangle is 45°, then

$$\frac{\text{Far side}}{\text{Near side}} = 1 \quad \text{or} \quad \frac{\text{Opposite side}}{\text{Adjacent side}} = 1$$

Right triangles that contained two 45° angles proved to be easy to analyze. However, just as the king predicted, soon every town wanted its own tree. We needed to calculate the heights of many different trees. The problem was that we could not always go to a point where the angle of elevation of the top of the tree was 45°. The very next day, we found we needed to calculate the height of a tree given this information. (See Figure 2-6.)

Figure 2-6

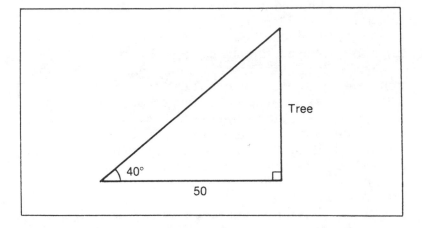

We found that there was a 40° angle formed by the ground and the line joining the top of the tree to our viewpoint 50 feet away from the tree. We called this angle the angle of interest and quickly determined that the opposite angle was 50°. Recordis drew a picture of the situation on his sketchpad. (See Figure 2-7.)

Figure 2-7

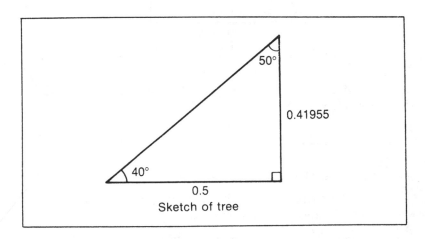

"This is hopeless!" Recordis moaned. "A right triangle with a 45° angle was easy, but there is no way to find the length of the far side in this case." Recordis took great pride in being able to accurately measure anything, so he began to worry that the others might think he had lost his touch. "There is one thing I can do, at least. I can measure the near side and the far side of the triangle in the little picture I just drew." He pulled out his most accurate ruler and found that the near side was 0.5 feet long and the far side was 0.41955 feet long.

(You might try doing this at home, but you will find it is very difficult to measure the length of the side as accurately as Recordis did.)

The professor had become intrigued by the ratio of the far side over the near side in the previous triangle we had investigated, so she suggested that we calculate the same ratio for this triangle:

Angle of interest = 40°

$$\frac{\text{Far side}}{\text{Near side}} = \frac{0.41955}{0.5} = 0.8391$$

"It is easy to calculate a ratio like that for a little triangle drawn on a piece of paper!" Recordis exclaimed. "If only this same relationship would be true for the big triangle formed by the tree and the ground!" (See Figure 2-8.)

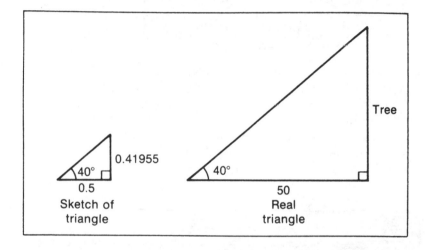

Figure 2-8

"When you think about it, those two triangles do look the same," the king said. "They both have exactly the same shape, even though the tree triangle is much bigger than the sketch of the triangle."

"Triangles with the same shape but different sizes are called *similar triangles*. You may read about them in my book if you don't remember them," the professor told Recordis. "Consider a pair of similar triangles. If one side of the big triangle is twice as long as its corresponding side on the little triangle, then *all* sides of the big triangle must be twice as long as their corresponding sides on the little triangle. In general, the sides of a pair of similar triangles will all be in the same proportion. We can see that the near side in the little triangle is 0.5 feet, and the near side in the big

**Calculating
Heights with
Similar Triangles**

triangle is 50 feet. Therefore, each side in the big triangle is 100 times as long as its corresponding side in the little triangle."

	Little Triangle	Big Triangle
Near side	0.5	50
Far side	0.41955	41.955

"We have found the height of the tree!" the professor explained. "The tree is 41.955 feet high."

"We have also discovered another useful result that will help whenever we confront a right triangle that contains a 40° angle," the king said. "For the small triangle, we can calculate that

$$\frac{\text{Far side}}{\text{Near side}} = \frac{0.41955}{0.5} = 0.8391$$

"We will get the same result if we perform the same calculation for the big triangle:

$$\frac{\text{Far side}}{\text{Near side}} = \frac{41.955}{50} = 0.8391$$

"In fact, we would get the same result if we perform this calculation for any right triangle that contains a 40° angle. Every right triangle containing a 40° angle is similar to every other right triangle that contains a 40° angle. Therefore, the ratio far side/near side will be the same for all these triangles." The king made a formal proclamation.

In a right triangle, if the angle of interest is 40°, then

$$\frac{\text{Opposite side}}{\text{Adjacent side}} = 0.8391$$

(This result is only an approximation to the true result. The true result is a decimal fraction consisting of an endless list of digits that never repeat a pattern.)

As winter approached, there were many other preparations that needed to be made. Builder proceeded to make plans for the new ski jump. "The ramp will be

25 yards long," Builder explained. "We have decided that we want the ramp to rise at a 10° angle." (See Figure 2-9.) "However, I need to know how high the support tower must be."

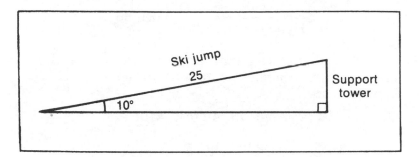

"Another right triangle problem!" the professor said excitedly. However, her excitement faded when she suddenly realized that we had not solved a problem like this before.

"In this problem we don't know either the near side or the far side," Recordis complained.

"We do know the hypotenuse, though," the king said encouragingly.

"And we know that the angle of interest is 10°," the professor said.

We drew a little triangle similar to the ski jump triangle on a paper. (See Figure 2-10.) We used the letter y to represent the length of the far side. Our triangle had a hypotenuse of length 0.25 yards, so each side on the big triangle was 100 times longer than its corresponding side on the little triangle. "Now we have to measure the length of y very accurately."

Figure 2-10

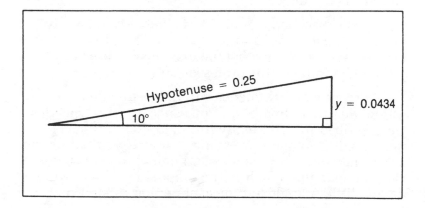

We found that y measured 0.0434 yards. Therefore, the height of the support tower on the big triangle must be 100 times that height, or 4.34 yards.

"Now we have enough information to calculate the ratio of the far side over the hypotenuse when the angle of interest is 10°," the professor noted. We calculated

$$\frac{\text{Far side}}{\text{Hypotenuse}} = \frac{4.34}{25} = 0.1736$$

"We should save that result in case we confront any more right triangles containing 10° angles," Recordis said.

$$\text{Angle of interest} = 10°$$

$$\frac{\text{Opposite side}}{\text{Hypotenuse}} = 0.1736$$

We went to lunch at Joe's Cafe, where we happened to see the Royal Astronomer sitting glumly at a corner table. The astronomer had been up all night puzzling about a difficult problem. "I have never been able to measure the distance to a star!" he sobbed. "The stars are so far away that I have not yet been able to figure out a way to measure the distance. I fear that our knowledge of the universe will remain quite limited unless we are able to solve this very difficult problem."

"I am sure you will discover something," Recordis said encouragingly.

The Shifting Star

"Even worse, I am now finding problems with my equipment," the astronomer continued mournfully. "Just last night I was observing a star that I had last observed exactly 6 months ago. I remember the night well . . . and I found that my equipment measured a different position for the star! Mind you, the discrepancy was very slight. It was only 0.8 seconds. But I like to be exactly precise, and even that much of an error is too large."

"But 1 second $= 1/3600° = 0.000278° = 2.78 \times 10^{-4}$ degrees, so 0.8 seconds $= 0.000222° = 2.22 \times 10^{-4}$ degrees. When you say that the discrepancy is small, you aren't kidding," the professor said.

Recordis tried to cheer up the despondent astronomer by telling him of our success with triangles. He described the problem with the trees and the ski ramp, and concluded by saying, "Now, if you tell us the size of just one of the angles and one of the sides, we can calculate the length of the other sides. We need to draw a little triangle and measure either this ratio,

$$\frac{\text{Far side}}{\text{Near side}}$$

or this ratio,

The astronomer listened politely while Recordis began to draw a picture of a right triangle. Suddenly, the astronomer leapt to his feet. "I have it!" he cried. "It's obvious now why the star shifted position! Not only that, I know how to calculate the distance to the star!" He excitedly drew a quick diagram. (See Figure 2-11.) "I had completely forgotten an obvious fact—during the course of a year the Earth moves about the sun!"

"Six months ago, the Earth was on one side of its orbit. The star appeared to be in the direction shown here. However, since then the Earth has moved to the other side of its orbit. In that situation, then, of course the position of the star as seen against the more distant background stars must have changed slightly. And in

The Distance to the Star

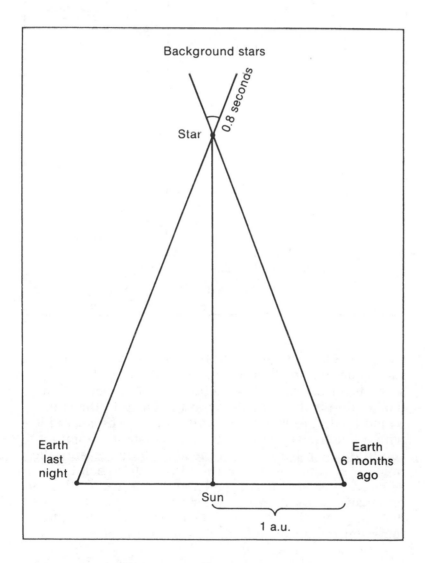

Figure 2-11

this case, we know that the total discrepancy was an angle of 0.8 seconds = 0.000222°."

The astronomer quickly drew the right triangle formed by the sun, the star, and the position of the Earth last night. (See Figure 2-12.)

Figure 2-12

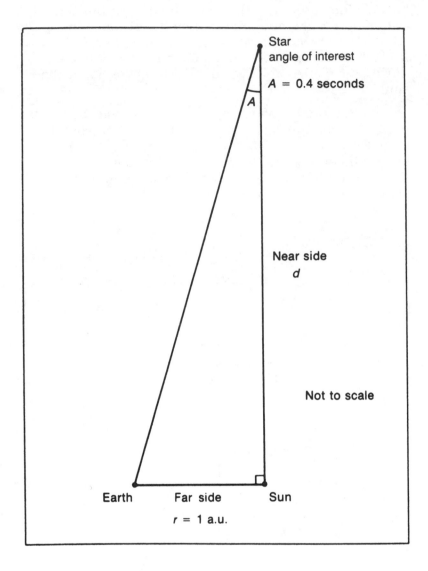

"Look at matters from the point of view of the star. Then our angle of interest is one half of 0.8 seconds, or 0.4 seconds = 0.000111°. That means that the far side is the distance from the Earth to the sun, which I call 1 astronomical unit (1 a.u.). The near side, with a length we don't know, is the distance from our sun to the star. Now, all you need to do is measure the ratio far side/near side for a right triangle when the angle of interest is 0.4 seconds, and then we shall have our answer." We used the letter r to represent the distance from the Earth to the sun and the letter d to represent the distance from the sun to the star.

Builder looked aghast when the problem was explained to him. "It will require extreme precision to draw a triangle with an angle that small," he cried. However, he pulled out his best etching equipment and drew a right triangle with the angle of interest equal to 0.4 seconds, the opposite angle equal to 89° 59 minutes 59.6 seconds, and the near side equal to 1 mile. Then he carefully measured the far side and found it equal to 0.0000019393 miles. (It would be incredibly difficult for you to try to draw the same triangle that Builder drew, and so we definitely need to find an easier way to solve this type of problem.)

Recordis wrote down that result:

Angle of interest = 0.4 seconds

$$\frac{\text{Far side}}{\text{Near side}} = 0.0000019393$$

"This ratio will hold for any right triangle with a 0.4 second angle, including the big triangle out in space," the professor said. Therefore,

$$\frac{r}{d} = 0.0000019393$$

From this formula we could calculate

$$d = \frac{r}{0.0000019393}$$

Since $r = 1$ a.u., we could calculate

$$d = \frac{1}{0.0000019393}$$

The approximate result was

$$d = 516,000 \text{ a.u.}$$

The astronomer jumped for joy. "We now have the answer to the problem that has been eluding us for years and years! We know that this star is 516,000 times farther away than the sun is. This method of triangles will be very useful—we will be able to find the distances to many different stars this way."

The professor was beginning to see a pattern in all these problems. "It seems to me that the nature of triangles is more subtle than we realize," she said.

- See the appendix on significant digits for a note on how to round off the result of a measurement.

- Recordis wanted the answer for the distance to the star expressed in terms of a unit he understood better. The astronomer told him that since 1 a.u. = 93,000,000 miles, we could write the distance like this:

Notes to Chapter 2

$$d = 516{,}000 \text{ a.u.} = 48{,}000{,}000{,}000 \text{ miles}$$

$$= 4.8 \times 10^{13} \text{ miles}$$

When Recordis saw the size of that number he was sorry that he had asked. The astronomer told him that he usually used *light years* to measure very large distances. A light year is the distance light can travel in 1 year. One light year equals $5{,}900{,}000{,}000{,}000$ miles $= 5.9 \times 10^{12}$ miles. Then we could express the distance to the star in light years:

$$d = \frac{4.8 \times 10^{13}}{5.9 \times 10^{12}} = 8.14 \text{ light years}$$

● The distances to stars were first measured using the method described in this chapter. Friedrich Bessel measured the distance to a star known as 61 Cygni in 1838. He found that the star had shifted by an angle of 0.30 seconds, and he calculated that the distance to the star must be 11 light years. This method of finding the distance to stars is known as *trigonometric parallax*. The distances to many other stars have been found by trigonometric parallax. The nearest star is a small companion of Alpha Centauri, which has a parallax shift of about 0.8 seconds and a distance of 4.3 light years. Note that closer stars have a larger parallax shift. However, when stars are farther away than 150 light years, the shifts are too small to be measured.

Exercises

For Exercises 1 to 11, fill in the missing values in the table for right triangles. Use the values for the ratios that are given in the chapter.

	Angle of interest	Adjacent side	Opposite side	Hypotenuse
1.	45°	16	—	—
2.	45°	—	—	$\sqrt{8}$
3.	—	1	—	$\sqrt{2}$
4.	40°	10	—	—
5.	40°	—	16.5	—
6.	40°	—	—	14.5
7.	—	20	16.782	—
8.	10°	16.54	—	—
9.	10°	—	—	0.1777
10.	10°	—	17.633	—
11.	—	567.13	100	—

12. Consider a right triangle with angle of interest A. Suppose $t = $ (opposite side)/(adjacent side) for this triangle. Suppose we now look at things from the point of view of the other acute angle. Show that $t = $ (adjacent side)/(opposite side) when seen from that angle.

RAGING RIVER

x^2

x

y

3

Trigonometric Functions: sin, cos, and tan

We found many applications for our triangle-solving methods in the next few days. We calculated the heights of many more trees, and surveyors and navigators found uses for the new methods. The astronomer quickly made plans to measure the distances to several important stars.

However, soon problems set in. A backlog developed of triangles waiting to be solved. It seemed as if everybody in the city was coming to Builder's desk and telling him the known parts of a triangle that needed to be solved. Then, Builder carefully drew the picture and measured the length of the unknown sides and reported back to the customer. However, people brought in triangles faster than Builder could draw them. Finally, Builder pleaded for help before the royal court. "There must be a better way," he cried. "The worst part is that sometimes people bring in triangles

Too Many Triangles

that I have already drawn before, but I have to draw them again each time."

"If you think you have problems now, just wait!" an ominous voice cried. In the next instant there stood before us a terrifying apparition in a deep black cape.

The Gremlin's Vile Threat

"The gremlin!" Recordis cried in terror. He instantly recognized the archenemy of the people of Carmorra—the Spirit of Hopelessness and Impossibility! His goal was to disrupt the entire learning process.

"We have defeated you each time we were confronted by one of your previous challenges!" the king said defiantly. "You claimed that we could not learn algebra, but we succeeded anyway."

The gremlin only laughed his cackling laugh. He pointed behind him to a huge pile of unassembled steel girders. "I dare you to construct a bridge over Raging River," he challenged us. He held out his cape and we saw a picture of a carefully crafted, arched bridge. In spite of the danger we could not help marveling at the graceful symmetry and balance of the bridge design. Upon looking closer we could see that the bridge was made up of many steel bars arranged to form hundreds of triangles.

"This is what the finished product would look like in the extremely unlikely event that you should succeed. However, you will find that your inability to solve triangles will be your downfall," the gremlin cackled. "When you fail, I shall take over and become king of Carmorra!" The gremlin vanished from sight, but his laughter still rang in our ears.

Recordis began to tremble, but Builder looked confidently at the pile of steel parts. "This job will be a piece of cake," he said. "We only have one problem—we must find a faster way to solve triangles."

Recordis panicked. "We don't know a faster way to solve triangles."

We thought about this problem for hours, but we had no success.

"Let's pass a law making all triangles illegal except for right triangles with 45° angles," Recordis suggested. "We know how to solve those." He turned to a page in his record book where he had recorded this result:

$$\text{Angle of interest} = 45°$$

$$\frac{\text{Far side}}{\text{Near side}} = 1$$

"We can also calculate the ratio far side/ hypotenuse for this type of triangle," the professor said. She drew a diagram. (See Figure 3-1.)

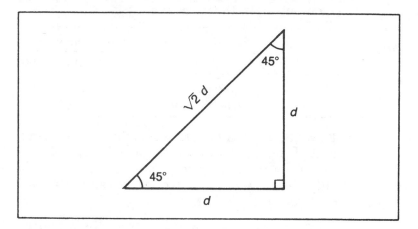

Figure 3-1

"If we let d represent the length of the two short sides, then we know from the Pythagorean theorem that the length of the hypotenuse must be $\sqrt{d^2 + d^2} = \sqrt{2d^2} = \sqrt{2}\ \sqrt{d^2} = \sqrt{2}d$."

Therefore,

$$\frac{\text{Far side}}{\text{Hypotenuse}} = \frac{d}{\sqrt{2}d} = \frac{1}{\sqrt{2}}$$

(We calculated a decimal approximation for this ratio: $1/\sqrt{2} = 0.7071$.)

We added this result to the table:

Angle of interest	Far side / Near side	Far side / Hypotenuse
45	1	0.7071

"That still doesn't help much," the professor said sadly. "Most triangles that we must deal with are not right triangles with two 45° angles."

"But there are two more types of triangles that we solved for," the king said. "We know how to solve a right triangle if it contains a 10° angle or a 40° angle."

We added these results to our table (see Chapter 2):

Angle of interest	Far side / Near side	Far side / Hypotenuse
40	0.8391	
10		0.1736

"By using the Pythagorean theorem we can fill in the two missing elements in this table," the professor said. (See Exercises 62 and 63.)

We came up with these results.

Angle of interest	Far side / Near side	Far side / Hypotenuse
40	0.8391	0.6428
10	0.1763	0.1736

"Let's write down a general procedure for solving for the length of the unknown side in a right triangle when we know these ratios," the professor said systematically. "First, let's pick two letters, such as *x* and *y*, to represent the two ratios."

"We already use *x* and *y* to represent the *x* and *y* axes," Recordis objected.

"All right, we'll use a couple of different letters— let's say, *s* and *t*," the professor agreed. She made these definitions.

Suppose *A* is the angle of interest in a right triangle. Then we will define

$$s = \frac{\text{far side}}{\text{hypotenuse}} \qquad t = \frac{\text{far side}}{\text{near side}}$$

(Note that these ratios will be the same for all triangles when the angle of interest is *A*, regardless of the size of the triangle.)

"Now, here's the general procedure," the professor said. "We'll use *h* to represent the length of the hypotenuse, *x* to represent the length of the near side, and *y* to represent the length of the far side. (See Figure 3-2.) Then,

Figure 3-2

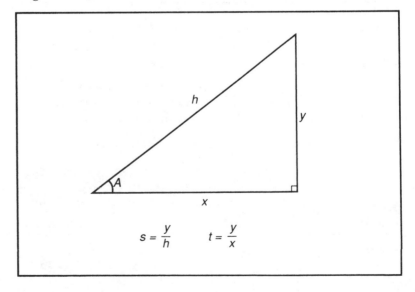

$$s = \frac{y}{h} \qquad t = \frac{y}{x}$$

1. If you know the near side and you would like to know the far side, use

$$y = tx$$

2. If you know the far side and you would like to know the near side, use

$$x = \frac{y}{t}$$

3. If you know the far side and you would like to know the hypotenuse, use

$$h = \frac{y}{s}$$

4. If you know the hypotenuse and you would like to know the far side, use

$$y = sh$$

5. If you know the far side and the near side and you would like to know the hypotenuse, use

$$h = \sqrt{x^2 + y^2}$$

(This is the Pythagorean theorem.)

"We now know these three types of triangles backward and forward," Builder said. "However, there are still many other types of triangles out there."

"Let's see if we can extend our table to cover other types of triangles," the professor said with sudden inspiration.

We realized that a right triangle with a 40° angle also contained a 50° angle. Likewise, a right triangle with a 10° angle also contained an 80° angle. So we were able to extend our table a little bit.

Angle of interest	$t = \dfrac{\text{Far side}}{\text{Near side}}$	$s = \dfrac{\text{Far side}}{\text{Hypotenuse}}$
10	0.1763	0.1736
40	0.8391	0.6428
45	1.0000	0.7071
50	1.1918	0.7660
80	5.6713	0.9848

(See Exercise 64.)

Before we could make any more progress, we were interrupted by a visit from Mrs. O'Reilly, the owner and manager of the Carmorra Beachfront Hotel, who came to ask Builder's help designing a holiday lighting display. "We would like a frame of lights forming a perfect equilateral triangle, supported by a post in the middle." (See Figure 3-3.)

Figure 3-3

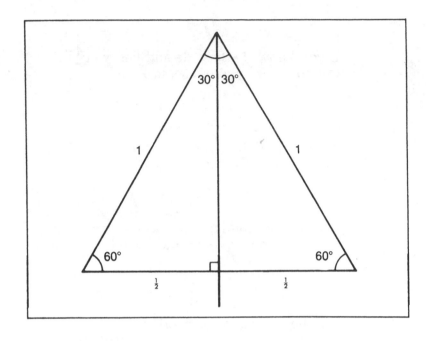

"We would like each side of the equilateral triangle to be exactly 1 unit long," Mrs. O'Reilly explained. "And make sure that the post forms a perfect right angle with the base of the triangle."

"That means that the post cuts the equilateral triangle into two right triangles!" the professor suddenly realized. "We can tell that the hypotenuse of each right triangle is 1 unit long and the shortest side of each right triangle has length $\frac{1}{2}$. And each right triangle must contain a 60° angle." (See Figure 3-4.)

Figure 3-4

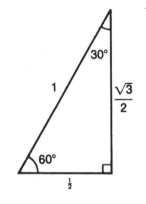

"The other angle in the right triangle must measure 30°," the king added.

"And we can use the Pythagorean theorem to calculate the length of the other side," Recordis added helpfully. We found that the length of the side opposite the 60° angle must be

$$\sqrt{1-\left(\frac{1}{2}\right)^2} = \sqrt{\frac{3}{4}} = \frac{\sqrt{3}}{2}$$

"Now we can calculate the two ratios when the angle of interest is 60°," the professor said.

$$t = \frac{\text{far side}}{\text{near side}} = \frac{\frac{\sqrt{3}}{2}}{\frac{1}{2}} = \sqrt{3} = 1.7321$$

$$s = \frac{\text{far side}}{\text{hypotenuse}} = \frac{\frac{\sqrt{3}}{2}}{1} = \frac{\sqrt{3}}{2} = 0.8660$$

"We may as well calculate the two ratios when the angle of interest is 30 degrees," Recordis said.

$$t = \frac{\text{far side}}{\text{near side}} = \frac{\frac{1}{2}}{\frac{\sqrt{3}}{2}} = \frac{1}{\sqrt{3}} = 0.5774$$

$$s = \frac{\text{far side}}{\text{hypotenuse}} = \frac{\frac{1}{2}}{1} = \frac{1}{2} = 0.5000$$

"That's a regular old rational number!" Recordis said delightedly. "I never did like irrational numbers very much—particularly irrational numbers involving square root signs."

We added these results to the table.

Angle of interest	$t = \dfrac{\text{Far side}}{\text{Near side}}$	$s = \dfrac{\text{Far side}}{\text{Hypotenuse}}$
10	0.1763	0.1736
30	0.5774	0.5000
40	0.8391	0.6428
45	1.0000	0.7071
50	1.1918	0.7660
60	1.7321	0.8660
80	5.6713	0.9848

"I'm beginning to get an idea," the king mused as he stared at that table. But, at that moment we were interrupted by the arrival of a tall gentleman carrying a strange large contraption.

"Allow me to introduce myself," he said. "My name is Alexanderman Trigonometeris, and I have just the item to help you solve all your holiday decorating needs—the Adjustable Triangle." (See Figure 3-5.)

The Decorative Adjustable Triangle

"I know him," Builder whispered to me. "He has presented me with many unusual inventions before. He tries very hard, but somehow his ideas never turn out to be useful."

Trigonometeris described his device.

"You may form whatever shape of right triangle you like," he said. "To do that, you adjust this bar that represents the hypotenuse. The hypotenuse is designed to rotate within a circle of radius 1—so all your right

Figure 3-5

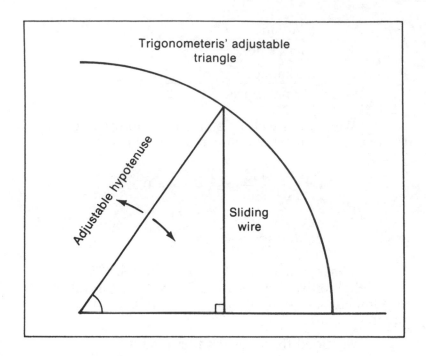

triangles will have a hypotenuse of length 1. The sliding wire that forms the vertical side is carefully designed so that it is always at a right angle to the horizontal side." Trigonometeris demonstrated how his device could form a triangle with a 30° angle. Then he adjusted the bar and formed a new triangle with a 45° angle.

We were all intrigued by his machine, but finally the king told him sadly, "I am afraid we do not need any more decorations this year."

Trigonometeris blinked back a tear. "Perhaps I could interest you in one of my other devices. . . ." He began to describe some of his other inventions.

Recordis cut him off. "We have serious business to conduct," he said. "The very survival of the kingdom is at stake. There is no way you could help us unless you could measure the ratios far side/near side and far side/hypotenuse for all possible right triangles."

"I'll find a way to do that," Trigonometeris bluffed, trying to conceal the fact that he did not understand exactly what Recordis meant. Stalling for time, he said, "Let's form a right triangle with a 5° angle." (See Figure 3-6.)

"We can measure the length of the near side and the far side for that triangle," the king said.

$$\text{Near side} = 0.9962$$

$$\text{Far side} = 0.0872$$

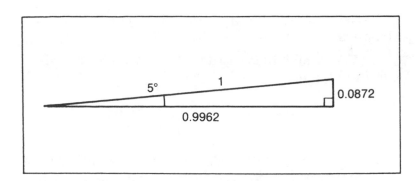

Figure 3-6

"Is *that* all you needed?" Trigonometeris exclaimed when he saw this result. "My triangles can do this easily. Just set the triangle to whatever angle you want, and then measure the sides!"

"We *can* use these triangles!" the professor said excitedly. "We can measure these ratios for all possible right triangles. It will be tedious, but when we're done, we can write down the results, and then we won't have to perform the same measurements again."

The professor started a table.

Angle	Near side	Far side	Far side / Hypotenuse	Far side / Near side
5°	0.9962	0.0872	0.0872	0.0875

"Hold everything!" Recordis said. "We must come up with names for these ratios before we go any further! I'm not going to write 'far side/hypotenuse' each time!" Recordis's job involved a lot of writing, so he frequently suffered from writer's cramp. He was always looking for ways to reduce the amount of writing that he must do. Indeed, one of our main motivations for developing the entire subject of algebra had been so we could express complicated mathematical problems using concise notation. Naturally, algebra became Recordis's favorite subject.

There was a terrible argument over the names for these ratios. Everyone wanted the ratios named after themselves. We were finally interrupted when Pal spilled four of his letter blocks on the Main Conference Room floor. They spelled the word *sine*. The king took decisive action to settle the argument. "We will call this ratio the *sine* ratio," he decreed.

The Sine Ratio

$$\text{Sine} = \text{ratio of } \frac{\text{opposite side}}{\text{hypotenuse}}$$

"The aerodynamic properties of letter blocks would make a fascinating study," the professor said.

"Will you stick to the subject!" Recordis cried. "You always go off on tangents!"

"Very well," the king declared. "We will call the other ratio the *tangent* ratio."

The Tangent Ratio

$$\text{Tangent} = \text{ratio of } \frac{\text{opposite side}}{\text{adjacent side}}$$

The names are too long," Recordis complained.

"We will use three-letter abbreviations for each ratio," Trigonometeris said. "We will use 'sin' to stand for sine, and we will use 'tan' to stand for tangent."

$$\text{sin} = \text{ratio of } \frac{\text{opposite side}}{\text{hypotenuse}}$$

$$\text{tan} = \text{ratio of } \frac{\text{opposite side}}{\text{adjacent side}}$$

(Note that the use of the name *sin* does not mean that this particular trigonometric ratio is morally degenerate. In trigonometry, the word *sin* is pronounced with a long *i*, as in *sign*.)

"But the word *sin* does not represent one particular value," the professor objected. "It could represent many possible values, depending on the value of the angle of interest. For example, we found that the sin ratio is 0.6428 if the angle is 40°, but the sin ratio is 0.1736 if the angle is 10°."

"We will write the angle of interest after the sin or the tan," Trigonometeris said. He was desperately trying to convince us that his triangles would be valuable. "We can write it like this:"

$$\sin 10° = 0.1736$$

$$\sin 40° = 0.6428$$

Functions

"This is what we call a *function*," the professor said. "We learned about functions when we studied algebra. A function converts one number into another number according to a rule. In our case, the sin function is a function that converts a number representing an angle into the sine ratio itself."

"Let's use the letter *A* to represent the angle of interest," Trigonometeris said. "Then we can calculate the sin function like this (Figure 3-7):

$$\sin A = \frac{y}{h}$$

"While we're measuring the length of the far side (*y*), we may as well measure the length of the near side

(x)," the king said. "Then we can calculate the tangent function as well:

$$\tan A = \frac{y}{x}$$

"For completeness, we should think of a special name for the ratio of the near side over the hypotenuse," the professor suggested.

"We'll call that the *cosine*," Trigonometeris said. (We later found out that he had a very good reason for using this name.)

$$\cos A = \frac{\text{adjacent side}}{\text{hypotenuse}}$$

$$= \frac{x}{h}$$

(We used cos as an abbreviation for cosine.)

The king issued a proclamation.

Figure 3-7

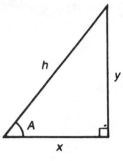

$$\sin A = \frac{y}{h}$$

$$\cos A = \frac{x}{h}$$

$$\tan A = \frac{y}{x}$$

Definition of Trigonometric Functions

Definition of the Sine, Cosine, and Tangent Functions

Draw a right triangle. Pick one of the non-right angles to be the angle of interest (call that angle A). Then,

$$\sin A = \frac{\text{opposite side}}{\text{hypotenuse}}$$

$$\cos A = \frac{\text{adjacent side}}{\text{hypotenuse}}$$

$$\tan A = \frac{\text{opposite side}}{\text{adjacent side}}$$

Let x represent the length of the adjacent side, y represent the length of the opposite side, and h represent the length of the hypotenuse. Then

$$\sin A = \frac{y}{h}$$

$$\cos A = \frac{x}{h}$$

$$\tan A = \frac{y}{x}$$

"We have already found some values for these functions," the king said. "For example, $\sin 30° = 1/2$; $\sin 60° = \sqrt{3}/2$, and $\sin 45° = 1/\sqrt{2}$."

"I see one obvious property," the professor said. "Since both y and x will always be less than h, it follows that $\sin A$ and $\cos A$ must always be less than 1 no matter what the value of A."

"But the value of the tangent function could be just about anything," the king said. "There is just as good a chance that x will be greater than y as there is that y will be greater than x."

"I see two more useful formulas that we can write," Trigonometeris said.

$$x = h\cos A$$

$$y = h\sin A$$

"These formulas follow directly from the definition of the functions."

"I think we're onto something big!" the professor said excitedly. "I think that these formulas will be very useful. We are starting a whole new subject."

"We would be honored if you could stay with us and work with us," the king told Trigonometeris. "We will call this new subject *trigonometry* in your honor."

Tears of joy glistened in Trigonometeris's eyes. At last he had found his calling in life. "We will give you the title of the Royal Keeper of the Triangles," the king continued.

That evening Trigonometeris used his triangles to measure the sine, cosine, and tangent of many different angles. He made a giant table containing these results. However, we decided it would be too awkward to keep looking at the table each time we needed one of these values, and so Builder built a device to allow us to conveniently find these numbers. "I call it a calculator," he said, and he showed us how it works. "To find the sine of 60 degrees, press the '6' and '0' keys. The display shows the number 60. Next, press the 'sin' key. The screen displays the answer: 0.866025403784. This is not the exact value of $\sin 60$ degrees, but it is an approximate value accurate to 12 decimal places." Trigonometeris could not contain his excitement as he realized how much easier it would now be to calculate the values of trigonometric functions.

(You should check your calculator's instruction book to learn the specific procedure for calculating trigonometric functions on your particular calculator. In the method described above, you enter the number 30 first and then press the sine key. This method works for many traditional calculators, and it also works for those that use reverse Polish notation. However, some newer calculators use a notation that is more like a computer programming language; you enter the sine function first and then enter the number 30. These calculators often let you use parentheses; for example, enter "sin(30)" to find the sine of 30.

(In Chapter 5 we learn that angles can be measured with *radian measure*, which is different from the degree measure we are using here. Many calculators allow you to use either radian or degree measure when calculating the value of trigonometric functions, but you must make sure that the calculator is set for the correct mode. Take a moment to learn how your calculator switches between radian mode and degree mode and for now make sure that it is set in degree mode. If you calculate sin 30 and get the result 0.5, you are in degree mode. If you calculate sin 30 and get the result −0.988, you are in radian mode.)

"Don't celebrate too quickly," Recordis cautioned. He was not sure that he liked this new subject very much because it involved so many strange names. "It is now up to Builder to save the kingdom by building the bridge."

Notes to CHAPTER 3

- A *function* in mathematics converts one number into another number according to a rule. For example, the function $f(x) = 2x$ means that the output number will always be equal to the input number multiplied by 2. The function $g(x) = x^2$ means that the output number will be equal to the input number raised to the second power—in other words, multiplied by itself. The input number to a function is called the *argument* or the *independent variable*. The output number from the function is called the *dependent variable*. In function notation, the name of the function is written first, followed by the argument enclosed in parentheses. For example, the expression

$$f(10) = 20$$

means that the name of the function is f, the argument is 10, and the output number is 20. The expression

$$\sin(30°) = \frac{1}{2}$$

means that the name of the function is sin, the argument is 30°, and the output number is $\frac{1}{2}$. If you need to review function notation, see a book on algebra.

- An *integer* is a whole number, such as 2 or 116 or 2117, or the negative of a whole number. We can see that the results for trigonometric functions are usually not integers. A *rational number* is a number that can be expressed as the ratio of two integers, such as $\frac{1}{2}$ or $\frac{2}{3}$ or $\frac{10}{11}$. For example, $\sin(30°) = \frac{1}{2}$, which is a rational number. A rational number can be expressed as a decimal fraction that either has a finite number of digits (such as $\frac{1}{2} = 0.5$, $\frac{1}{4} = 0.25$, or $\frac{5}{8} = 0.625$) or else consists of digits that endlessly repeat the same pattern (such as $\frac{1}{3} = 0.3333\ldots$; $\frac{1}{7} = 0.142857142857142857\ldots$; or $\frac{15}{11} = 1.36363636\ldots$). However, we have found that $\sin(45°) = 1/\sqrt{2}$, which is not a rational number. It is impossible to find two integers a and b such that $a/b = 1/\sqrt{2}$. This type of number is called an *irrational* number. An irrational number can be represented as a decimal fraction with digits that continue endlessly without ever repeating a pattern. For example, $1/\sqrt{2} = 0.7071067812\ldots$. The values of the trigonometric functions for most angles are irrational numbers. However, it is even worse than that. Even though $\sin(45°)$ is irrational, there is a simple formula for this number using a square root sign: $\sin(45°) = 1/\sqrt{2}$. The values of trigonometric functions for most angles cannot even be represented by a formula like this. The trigonometric function values for most angles are called *transcendental numbers*. A transcendental number is a special type of irrational number. For our purposes it is sufficient to know that you cannot find an expression of the form $y = p^q$ (where p and q are both rational numbers) if y is a transcendental number. Therefore, square roots and cube roots are not transcendental even though they are irrational. There are two very special transcendental numbers in mathematics: $\pi = 3.14159\ldots$ and $e = 2.71828\ldots$. Also, if you have studied logarithm functions you will have learned that the values of logarithms are usually transcendental numbers.

For practical purposes, the difference between transcendental numbers, irrational nontranscendental numbers, and rational, endlessly repeating numbers does not matter much. In each case you will represent the true value as a decimal approximation.

Exercises

1. You will need to memorize the definitions of the sine, cosine, and tangent functions. You should do that now.

2. You should know the exact values of the sine, cosine, and tangent functions for these special angles: 30°, 45°, and 60°. Make a table listing the values of those functions for those angles.

3. The very first evening he was at the palace, Trigonometeris discovered a very important relation. He found

$$\tan A = \frac{\sin A}{\cos A}$$

for any value of A. Use the definition of these three ratios to prove that this relation is true.

Find the values for the sine function, the cosine function, and the tangent function for the angles in Exercises 4 to 10. Use a calculator.

4.	10°	7.	76.6°
5.	15°	8.	16.4°
6.	33.4°		

9. 45°

10. 12°

11. Show that $\sin A = \cos(90° - A)$.

12. Suppose that you only had values of the trigonometric functions from 0 to 45°. How could you still calculate the values of the functions for the other angles?

For Exercises 13 to 24, fill in the missing values in the following table for right triangles.

	Angle of interest	Adjacent side	Opposite side	Hypotenuse
13.	30°	50	—	—
14.	30°	—	$16\sqrt{3}$	—
15.	30°	—	—	18
16.	60°	12	—	—
17.	60°	—	24	—
18.	60°	—	—	1
19.	10°	16.34	—	—
20.	10°	—	3.64	—
21.	35°	—	—	15.846
22.	42°	7.3	—	—
23.	47.5°	—	10.913	—
24.	58.4°	—	—	31.508
25.	45°	56	—	—
26.	45°	—	125.8	—
27.	45°	—	—	858
28.	2°	18	—	—

29.	2°	—	36.5	—
30.	2°	—	—	15.73
31.	87°	158	—	—
32.	87°	—	788	—

The angle of elevation of an object is the angle between the horizontal and the line connecting your position to the object (assuming that the object is above you). See Figure 3-8. Complete the following table for Exercises 33 to 38.

Figure 3-8

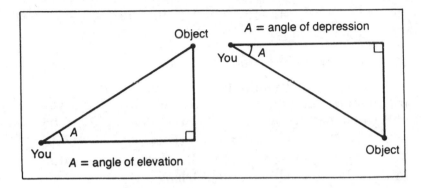

	Angle of elevation	Distance	Height
33.	20°	100	—
34.	20°	—	100
35.	40°	65	—
36.	40°	—	436
37.	75°	—	30
38.	75°	900	—

If you are looking at an object that is below you, you may calculate the angle of depression. See Figure 3-8. Complete the following table for Exercises 39 to 44.

	Angle of depression	Distance	Depth
39.	10°	36	—
40.	10°	—	245
41.	30°	1.74	—
42.	30°	—	26.45
43.	52°	1182	—
44.	53°	—	75.46

45. If you are given the size of an object and its angular size, derive a formula that tells you its distance. (See Figure 3-9).

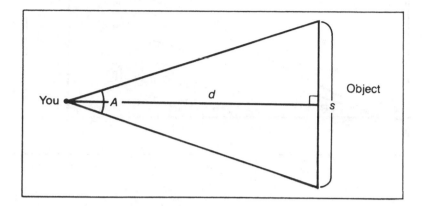

Figure 3-9

For Exercises 46 to 53, calculate the distance to the objects.

	Objects	Size	Angular size
46.	Mt. Rainier, seen from Seattle	2.7 miles	2.578°
47.	Width of Central Park, seen from Empire State Building	3000 feet	21.239°
48.	Earth, seen from moon	12,750 kilometers	1.9°
49.	Moon	3500 kilometers	0.522°
50.	Sun	864,000 miles	0.532°
51.	Saturn	75,000 miles	0.00537°
52.	Star Antares	5.5×10^8 miles	1.37×10^{-5} degrees
53.	Andromeda galaxy	130,000 light years	3.38°

54. Suppose you are standing an unknown distance away from a cliff of height h. You need to know the height t of a tower located on top of the cliff. You know that the angle of elevation of the bottom of the tower is B and the angle of elevation of the top of the tower is A. Derive a formula for the height of the tower. (See Figure 3-10).

Figure 3-10

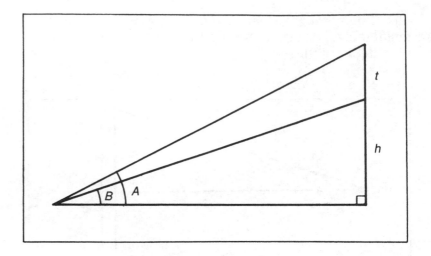

55. Suppose you need to calculate the height of a distant cliff. Unfortunately, you do not know the distance to the cliff. However, you have found the angle of elevation of the top of the cliff at one point is A_1 and the angle of elevation at another point that is d units farther away is A_2. (See Figure 3-11.) Derive a formula for the height of the cliff.

Figure 3-11

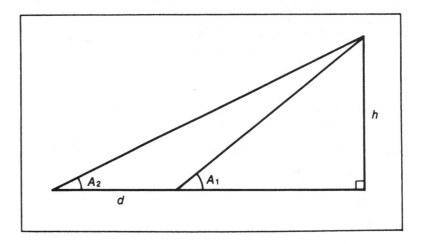

Suppose you need to calculate the height of a tower that is at the top of a distant cliff. You don't know the height of the cliff or the distance to the cliff, but you do know the angle of elevation of the top and bottom of the tower from two different points that are a distance d apart. (See Figure 3-12.) The following table gives you the value for A_1, A_2, B_1, B_2, and d for observations of several different towers on the tops of several different cliffs. For Exercises 56 to 61, calculate the height of each tower.

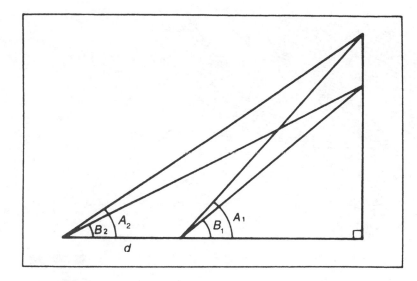

Figure 3-12

	A_1	B_1	A_2	B_2	d
56.	30°	28°	25°	23.2409°	30
57.	45°	40°	30°	25.8481°	20
58.	20°	19.4°	19°	18.4255°	100
59.	23°	20°	20°	17.3326°	50
60.	65°	40°	20°	8.1052°	1000
61.	18.6°	18.2°	18.0°	17.6112°	20

62. Before we invented the trigonometric functions, we found that (opposite side)/(adjacent side) = 0.8391 for a 40° angle contained in a right triangle. Use this fact to calculate the ratio (opposite side)/(hypotenuse) for a 40° angle. Do not use your calculator's sin button.

63. We found that (opposite side)/(hypotenuse) = 0.1736 for a 10° angle contained in a right triangle. Use this fact to calculate the ratio (opposite side)/(adjacent side) for a 10° angle. Do not use your calculator's tan button.

64. Show how you can derive values for sin 80° and tan 80° once you know the values for sin 10° and tan 10°. Show how you can derive the values for sin 50° and tan 50° once you know the values for sin 40° and tan 40°.

Applications of Trigonometric Functions

Trigonometeris eagerly reported for work at 7 the next morning. However, the rest of us did not arrive until 8:30, as usual. Trigonometeris was anxious to start work, but we had not decided exactly what duties should be attached to the office of the Royal Keeper of the Triangles.

"I'm sure we will find many applications for these new functions," Trigonometeris said excitedly.

The Balloon Ride

However, our main business for the moment was to travel to Raging River to observe Builder's progress with building the new bridge. We decided to take our propeller-driven helium-filled balloon. Recordis took his position as pilot while the rest of us, including Trigonometeris, climbed on board.

"Piloting a balloon requires careful navigation," Recordis said. "We must plot our course precisely. I

46

happen to know that, in order to get from Capital City to the bridge site, we must travel in a perfectly straight line in a direction that is 55° north of east." (See Figure 4-1.)

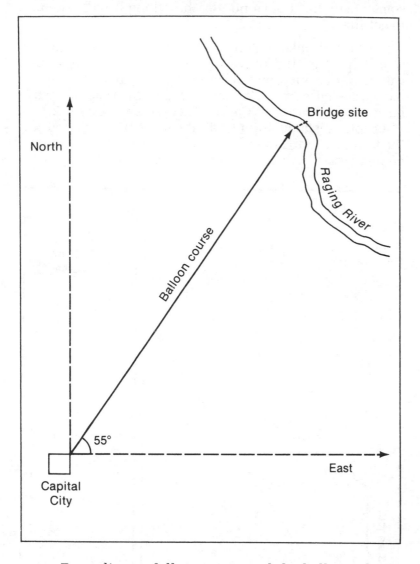

Figure 4-1

Recordis carefully maneuvered the balloon along our course. Fortunately, there was no wind, so it was easy to maintain a straight-line course. The balloon was carefully designed to travel at a constant speed of 15 miles per hour.

"I wonder when we shall cross over the Straight Arrow River," the professor said. "I like the view of that river from the air."

Recordis puzzled for a moment. "That is a very hard question," he said. "The Straight Arrow River flows in a perfect straight line from south to north, and we know that the river is 30 miles east of Capital City." (See Figure 4-2.) "If we were traveling directly east, then the answer would be obvious: We would have to travel 30 miles until we reached the river. Since we travel at 15 miles per hour, it would take us 2 hours. But we're not traveling directly east. We're traveling 55° north of east. Of course, we will still cross the river somewhere, but I have no idea how long it will take us to get there."

"It will take us longer than 2 hours," the King said helpfully. "We know that our position at any time can be represented by two numbers: the distance we have traveled *east* of Capital City, and the distance we have traveled *north* of Capital City." (See Figure 4-2.) "Our total distance from Capital City is increasing at the rate of 15 miles per hour, so the east distance must be increasing by less than 15 miles per hour."

Figure 4-2

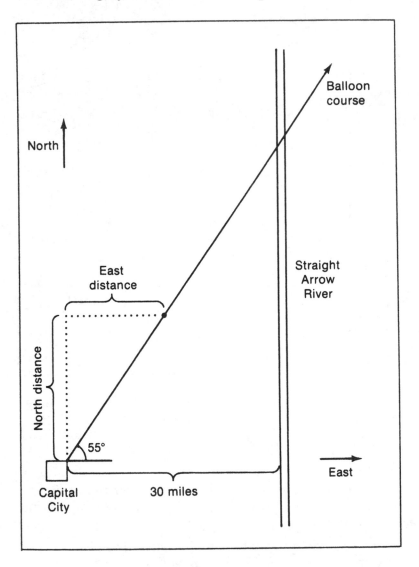

Recordis continued to concentrate on piloting. He was constantly checking a small card he held in his hand. "What is that little arrow on the card?" the professor asked with interest.

"I call that my *velocity vector*," Recordis said proudly. (See Figure 4-3.) "The velocity vector is an arrow that tells me about the course. You must be very careful when you draw a velocity vector. You must make sure that you draw both the *direction* and the *magnitude* correctly. The direction of the arrow points in the direction we are going. The magnitude (that's a long word that means "length") of the arrow is proportional to the velocity. In this case I have used a scale where a vector 1.5 inches long corresponds to a speed of 15 miles per hour."

Velocity Vectors

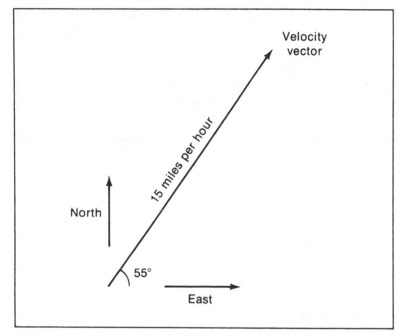

Figure 4-3

The professor suddenly became excited. "The vector that you have drawn represents the real vector formed by our course through the air. However, we can pretend that the real velocity vector is made up of two imaginary velocity vectors—one that points directly east, and one that points directly north." (See Figure 4-4.)

"I have enough trouble keeping track of one real vector," Recordis complained. "How am I going to be able to keep track of two imaginary vectors?"

"We'll call the vector that points east the *east component* of our motion, and we'll call the vector that points north the *north component*," the professor decided. "Now, if we could only calculate the length of the east component, we would know how fast we are moving east, and then we could calculate how long it will take us to reach the Straight Arrow River."

Suddenly Trigonometeris brightened. "We can use trigonometry!" he exclaimed. "We can see from the diagram that

Component Vectors

Figure 4-4

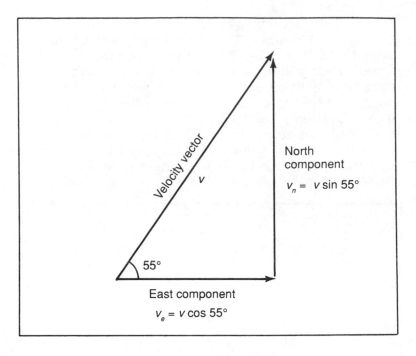

$$v_e/v = \cos 55, \quad \text{so} \quad v_e = v \cos 55$$

and

$$v_n/v = \sin 55, \quad \text{so} \quad v_n = v \sin 55$$

"where v_e stands for the east component of our velocity, v_n stands for the north component of our velocity, and v stands for the magnitude of our total velocity (in this case, $v = 15$ miles per hour)."

We found the value of $\cos 55$ with Trigonometeris's calculator (which he kept in a locked jeweled case about his neck):

$$\cos 55 = 0.5736$$

Therefore,

$$v_e = v \times 0.5736 = 15 \times 0.5736 = 8.604$$

"Therefore, the east component of our velocity is 8.604 miles per hour," Trigonometeris said. "That means that each hour we have traveled 8.604 miles farther east. Since the Straight Arrow River is 30 miles away, we will reach it in $30/8.604$ hours = 3.49 hours."

Just as we predicted, we crossed the river 3.49 hours after we had left, and the view was spectacular.

We decided that the method of breaking a velocity vector up into component vectors might be very useful for other types of problems as well.

Velocity Components

Suppose you have decided on two directions called the x direction and the y direction. (For example, the x direction might represent east and the y direction represent north, or the y direction might represent up and the x direction represent horizontal motion in a particular direction. These two directions must be at right angles to each other.) Then, suppose that v represents a velocity vector in a particular direction. (See Figure 4-5.) Then you may find the x component and the y component of the velocity vector according to the formulas

$$v_x = v \cos A$$

$$v_y = v \sin A$$

It was only an hour later when we arrived at the bridge site where Builder and Pal were already hard at work. "You are just in time to help me with some tricky problems," Builder said with relief. "I have a small rowboat I use to ferry supplies to the opposite shore. Pal always rows the boat straight across the river, and he always rows at a constant speed of 7 miles per hour.

The Off-Course River Boat

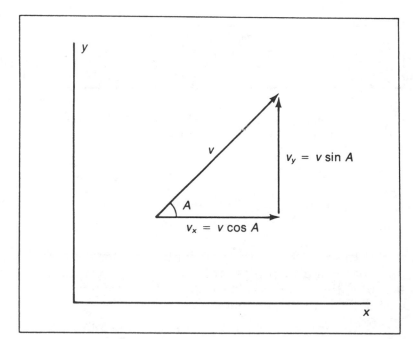

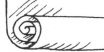

Figure 4-5

Figure 4-6

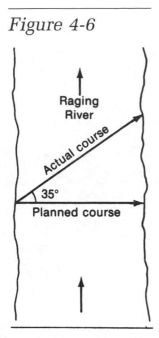

However, the boat always ends up traveling along a course that is off by 35°." (See Figure 4-6.) "I just can't figure it out."

"It almost looks as though the river current is pushing you off course," Recordis suggested.

Builder slapped his forehead. "How could I have been so stupid? It *is* the river current that makes the boat go off course. I wonder how fast the river is flowing?"

"Once again trigonometry will come to the rescue just in the nick of time," Trigonometeris said. "We need to draw three vectors: one vector representing the boat's course relative to the river (v_b), which points directly east; one vector representing the river current v_r, which points directly north; and one vector representing the boat's actual course." (See Figure 4-7.)

Figure 4-7

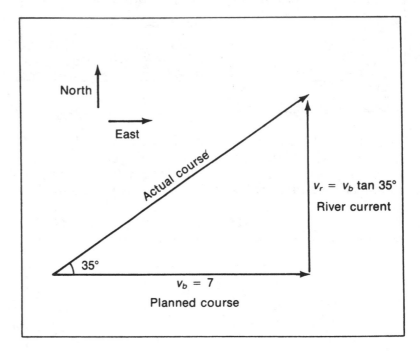

"We know that the length of the vector v_b is 7.0 miles per hour," the professor said. "Then,

$$\frac{v_r}{v_b} = \tan 35$$

$$v_r = v_b \tan 35$$

$$v_r = 7 \times \tan 35$$

To solve this with the calculator, enter the number 35 and then press the tan key. The result for tan 35 is displayed: 0.70020753821. (Be sure your calculator is in degree mode.)

Multiply this result by 7 to give: 4.90145276747.

However, the final result should be presented with only two significant digits: 4.9.

"So, the river is flowing at 4.9 miles per hour," Builder said. "That will be very useful to know."

We had three more interesting trigonometry adventures that day. However, these applications involve some tricky physics concepts. You may skip to the end of the chapter if you wish.

We rode across the river in the boat. When we reached the other side, Builder demonstrated his newest invention, a message-delivering system. He showed us several different slingshots constructed on the hillside. Each slingshot was tilted at a different angle. "When I need to send a message, I put the message inside a little capsule. Then I put the capsule inside one of the slingshots and send it in the direction it is supposed to go. Each slingshot is designed so that it fires the capsules at an initial velocity of precisely 35 meters per second. However, I need to know the *distance* that the capsule will travel before it hits the ground. The distance traveled naturally depends upon the angle at which the slingshot is aimed. If the slingshot is aimed too steeply upward, then the capsule will not travel very far because it wastes most of its motion going up. On the other hand, if the slingshot is not aimed very steeply, then the capsule will not travel very far either." (See Figure 4-8.)

"So, you want us to calculate the distance the capsule will travel as a function of the angle of tilt of the slingshot," the professor clarified the problem.

"To do that, we would need to know how gravity works," Recordis said. "All I know about gravity is that I get hit on the head if I fall asleep under an apple tree."

"I have discovered two formulas that describe the motion of the capsule in two special cases," Builder said helpfully. "If you aim the slingshot straight up, the time in seconds until the capsule lands is given by this formula:

$$t_{\text{land}} = \frac{2v_0}{g}$$

"In this formula, g represents a special number whose value is 9.8 and v_0 represents the initial velocity of the capsule, which is 35 meters per second. (I use t_{land} to represent the time until the object lands.) Although this formula is interesting, it is of no practical value for sending messages. If you shoot the capsule straight up all it does is come straight down again.

"I have also discovered a formula that describes what happens when you shoot the capsule off the cliff

Figure 4-8

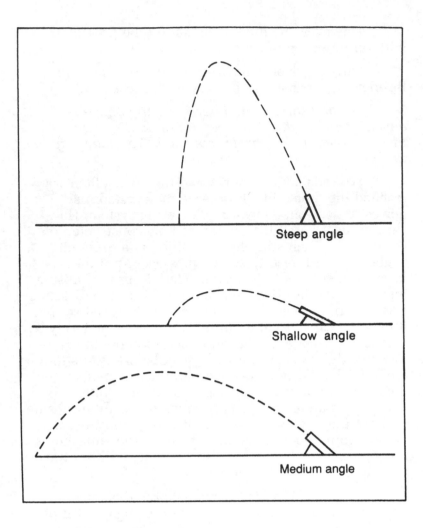

at a zero degree angle—in other words, you shoot the capsule horizontally. Then, the horizontal distance that the capsule has traveled at time t is given by this formula:

$$d = v_0 t$$

"So we can solve the problem if the capsule is shot horizontally or vertically—but not if the capsule is shot at any other angle," the professor said.

"Let's draw an initial velocity vector for the capsule," Trigonometeris said helpfully. "Then we can figure out a horizontal and vertical component of the initial velocity." We used v_0 to represent the vector of the initial velocity, A to represent the angle of tilt of the slingshot, v_h to represent the horizontal component of initial velocity, and v_v to represent the vertical component of initial velocity. (See Figure 4-9.)

"I know how we can use trigonometry to calculate the magnitude of the two components," the professor said.

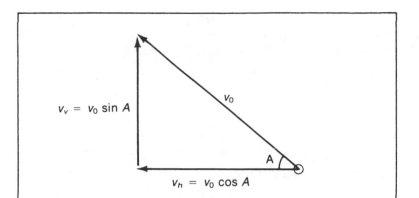

Figure 4-9

Vertical velocity component $= v_v = v_0 \sin A$

Horizontal velocity component $= v_h = v_0 \cos A$

"We know that $v_0 = 35.00$. Then, for example, if $A = 25°$ we can find from our calculator that $\sin 25° = 0.4226$ and $\cos 25° = 0.9063$. Therefore, $v_v = 35 \times 0.4226 = 14.79$ and $v_h = 35 \times 0.9063 = 31.72$.

"That doesn't help us solve the problem, though!" Recordis moaned. Try as we might, we could not figure out how to calculate the distance the capsule would travel. Finally, Recordis said, "I always say, when you are faced with a difficult problem that you can't solve, make up a new problem that you can solve. For example, let's suppose that we shot a capsule straight up with a velocity 14.79 (which is the vertical component of the velocity when the slingshot is tilted at a 25° angle). Then we can calculate the time until it hits the ground from the formula:

$$t_{\text{land}} = \frac{2v_v}{g} = 3.02 \text{ seconds}$$

"I know another problem we could solve," the king said. "Suppose that we shot a capsule horizontally with an initial speed of 31.72. Then we know from the formula $d = v_0 t$ that the distance that it would travel in 3.02 seconds would be 3.02×31.72 or 95.79 meters."

While we were working on this problem Pal came by playing with his beachball. He was throwing the beachball up in the air at different angles. The professor decided to carefully monitor the motion of the beachball, and she discovered an amazing fact.

"The formula $t_{\text{land}} = 2v_v/g$ still gives you the time until the object hits the ground, whether or not the object is launched straight up! And I discovered something else. The formula $d = v_h t$ still gives you the horizontal distance that the object has traveled from the starting point, whether or not the object is launched horizontally.

The king exclaimed, "Let's put these two formulas together!" The horizontal distance the object will travel before it hits the ground:

★**The Distance of Travel of the Capsule**

$$d = v_h t_{\text{land}}$$

$$d = v_h \frac{2v_v}{g}$$

Then,

$$d = \frac{2v_h v_v}{g}$$

"Let's use these trigonometry formulas," Trigonometeris suggested.

$$v_v = v_0 \sin A$$

$$v_h = v_0 \cos A$$

"Then we find

$$d = \frac{2(v_0 \sin A)(v_0 \cos A)}{g}$$

$$d = \frac{v_0^2}{g} 2 \sin A \cos A$$

"That's just the answer we want—it expresses the horizontal distance that the object will travel as a function of the angle of inclination." (We later found that we could write that formula like this: $d = (v_0^2/g) \sin 2A$. See Chapter 6.)

"Let's calculate some sample values," Builder said. He told us the angle of tilt for each slingshot.

We first calculated the result where the angle of launch was 10.0°. (Recall that the initial velocity is 35 meters per second.)

$$d = \frac{35^2}{9.8} \times 2 \times \sin 10 \times \cos 10$$

$$d = 250 \times 0.1736 \times 0.9848 = 42.8$$

We made a table for some other values:

Angle of launch	Distance traveled (meters)
10	42.8
20	80.3
30	108.3
40	123.1
45	125.0
50	123.1
60	108.3
70	80.3
80	42.8

★**The Slippery Slope**

"Aha! Just as I suspected," Builder said. "The capsule will travel the greatest distance if it is launched at an angle of 45°. However, I still have a problem with designing the approach road for the bridge. The road must travel through steep mountains. I need to figure out the steepest possible slope we can allow. Obviously, if the road is too steep, then cars will slide down the hill. I need to know the steepest allowable angle."

"What makes the cars slide?" the professor asked, intrigued.

"Everybody knows what makes something slide down a hill!" Recordis exclaimed.

"We understand that intuitively," the professor said. "But we should specify exactly what causes the motion."

"I use the term *force* to mean something that causes (or restrains) motion. For example, let's suppose that we have a car parked on the road," Builder said. (See Figure 4-10.) "Then there are three forces acting on the car: The force of gravity acts straight down, there is a constraint force that keeps the car from falling through the road, and there is a friction force that keeps the car from sliding down the road."

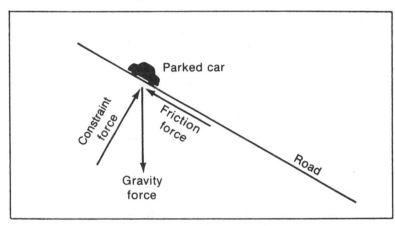

Figure 4-10

"We can represent each of these forces as a vector!" Trigonometeris realized. "For each force, we need to know the direction in which it points, and we need to know how strong the force is—in other words, the magnitude of the vector. (Although I personally have no idea what type of unit you use to measure the size of a force.) Let's use the letter A to represent the angle of tilt of the road. Then we can divide the gravity force into two components: the sideways component that pulls the car down the ramp, and the pressing component that keeps the car on the road." (See Figure 4-11. Use geometry to show that the two angles labeled A are equal.)

Trigonometeris let F_g represent the magnitude of the gravity force. Then we could calculate the magnitude of the two components:

Figure 4-11

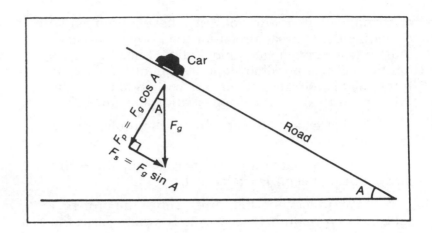

$$F_s = F_g \sin A \qquad \text{sliding component}$$

$$F_p = F_g \cos A \qquad \text{pressing component}$$

"The pressing component of the gravity force is exactly equal to the constraint force of the road," Builder said helpfully, "although they point in opposite directions. If the pressing component were greater than the constraint force, the road would collapse and the car would fall through."

"I see how we can tell whether the car will slide down the hill!" the professor said. "If the magnitude of the sliding component is greater than the magnitude of the frictional force, then the car will slide!"

"Everybody knows that!" Recordis said. "However, we don't have the faintest idea how to calculate the magnitude of the frictional force."

"The magnitude of the frictional force is proportional to the magnitude of the pressing force," Builder said helpfully.

$$F_f = (\text{some number}) \times F_p = (\text{some number}) \times F_g \cos A$$

"But how do we know what the value of that 'some number' is?" Recordis demanded.

★*Friction*

"That obviously depends on the road conditions," Builder said. "I call that quantity the *friction coefficient* (or f_c for short). If the road is icy, then the value of the friction coefficient is small, and the cars are much more likely to slide. Under normal circumstances, the value of the friction coefficient for this type of road is about 0.4."

We wrote out the equations:

$$F_s = F_g \sin A \qquad \text{sliding component of gravity}$$

$$F_f = 0.4\, F_g \cos A \qquad \text{friction force}$$

"So the car will slide if this inequality is true," the professor exclaimed.

$$F_s > F_f$$

or

$$F_g \sin A > 0.4 \, F_g \cos A$$

"We can cancel out that F_g, since it appears as a factor on both sides," Recordis said, cheering up a bit. Recordis found trigonometry to be very confusing, but he still loved to cancel things. The car will slide if

$$\sin A > 0.4 \cos A$$

"We can divide both sides by $\cos A$," the professor said. The car will slide if

$$\frac{\sin A}{\cos A} > 0.4$$

"We know that $\sin A/\cos A = \tan A$," Trigonometeris said, glad that one of the relations he had discovered the day before had come in handy. The car will slide if

$$\tan A > 0.4$$

Trigonometeris spent a long time punching numbers into his calculator, looking for an angle whose tangent was 0.4. He knew it had to be less than 30 degrees since $\tan 30° = 0.5$. He found $\tan 20° = 0.3640$, which is too small. He tried some more possibilities:

★*The Maximum Angle of Tilt*

Angle	Tangent
25°	0.4663 (too big)
22.5°	0.4142 (too big)
21.2°	0.3879 (too small)
21.9°	0.4020 (slightly too big)
21.7°	0.3979 (slightly too small)
21.8°	0.4000 (just right, for four decimal place accuracy)

"Therefore, if A is greater than 21.8°, then $\tan A$ is greater than 0.4, and the car will slide," Trigonometeris concluded.

(Fortunately, there is an easier way to find the size of an angle if you know the value of its tangent. There is an inverse function for tangent, which is known as the arctangent function. Look on your calculator for a button labeled "arctan," which can also be written as "atn" or "atan" or "tan⁻¹." When you enter the number 0.4 into your calculator and apply the arctangent function, you get a result of about 21.8 degrees. If you get a result of about 0.38, it means your calculator is in radian mode. Be sure to set it back to degree mode for now. Radian measure is discussed in

Chapter 5, and inverse trigonometric functions are discussed more in Chapter 10.)

"Just what I needed to know!" Builder said gratefully. "I must be very careful to design the road so that the steepest slope is not steeper than 21.8°."

★*The Merry-Go-Round Streamers*

Recordis eyes were bleary from doing this much work in one day. "Let's work on something fun, like the design for the new carnival merry-go-round we will build to celebrate the opening of the bridge," he said, "I would like the outer rim of the top to be decked with streamers with small balls at their ends. The radius of the outer rim is 10 meters. However, I still have a problem. I would like the streamers designed so that they hang outward, forming a 15° angle while the merry-go-round turns. (See Figure 4-12.) I don't know what the turning speed of the merry-go-round should be."

Figure 4-12

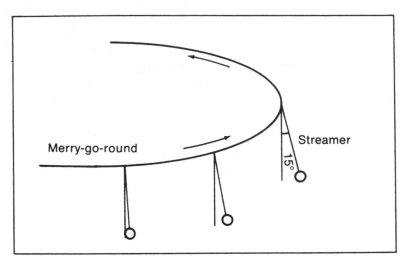

"Why do the streamers hang outward when the merry-go-round turns?" the professor asked. Like all brilliant theoreticians, she did not always have an intuitive understanding of practical matters.

"Everyone knows that when you turn something it seems to be pulled outward!" Recordis exclaimed. "Haven't you ever ridden on a wagon around a sharp curve? You feel like you are being pulled outward. When the merry-go-round is stopped, the streamers will hang straight down. When the merry-go-round starts moving faster, then the streamers will hang farther and farther out."

★*Centrifugal Force*

"I have a name for that type of force," Builder said. "I call it *centrifugal force*. It's not a real force, so I call it a *fictitious force*. Whenever something rides inside an object moving around in circles, it will seem to be feeling a centrifugal force pushing it outward. I have calculated that if f is the frequency of rotation (measured as the number of turns per second), r is the radius of the ride, and m is the mass of the ball, then the size of the centrifugal force is approximately

$$F_c = 39.48\, mrf^2$$

(The exact formula is $F_c = mr(2\pi f)^2$. Note that the force is greater if the ride turns faster.)

"Now it is a trigonometry problem!" Trigonometeris said. "We know that, while the merry-go-round is turning, a ball at the end of a streamer is acted upon by three forces: the force of gravity (F_g) pulling straight down, the centrifugal force pulling straight out, and the force of the streamer itself, which pulls the ball up at an angle." (See Figure 4-13.)

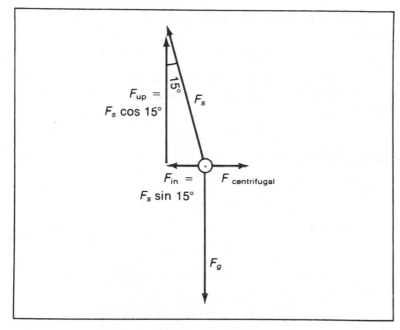

Figure 4-13

"These three forces must exactly cancel each other out, since we don't want the ball to be moving," the king said. ("Actually, of course, the ball will be moving if you stand on the ground to watch it. If you are riding on the merry-go-round itself, then it will not appear to you that the ball is moving.")

"We can divide the force of the streamer into two components: an upward component and an inward component," the professor said. Using A to stand for the angle of tilt of the string, we found

$$F_{up} = F_s \cos A \qquad \text{upward force}$$

$$F_{in} = F_s \sin A \qquad \text{inward force}$$

We wrote one equation stating that the upward force of the streamer was equal to the downward force of gravity:

$$F_s \cos A = F_g$$

and another equation stating that the inward force of the streamer was equal to the outward centrifugal force:

$$F_s \sin A = F_c = 39.48\, mrf^2$$

"Now we've reduced it to an algebra program!" Recordis said with relief. He rewrote the first equation to give us an expression for F_s:

$$F_s = \frac{F_g}{\cos A}$$

Then he substituted this expression for F_s into the second equation:

$$\frac{F_g}{\cos A} \sin A = 39.48 mrf^2$$

Trigonometeris reminded us that $\sin A / \cos A = \tan A$:

$$F_g \tan A = 39.48\, mrf^2$$

Builder told us that the magnitude of the gravity force depended on the mass of the balls according to the formula

$$F_g = mg$$

where g once again had the value $g = 9.8$.

Then we wrote

$$mg \tan A = 39.48\, mrf^2$$

Recordis gleefully canceled out the two m values:

$$g \tan A = 39.48\, rf^2$$

Then he filled in the values $g = 9.8$, $A = 15°$, and $r = 10$:

$$9.8 \tan 15 = (39.48)(10) f^2$$

$$0.00665 = f^2$$

$$f = 0.082 \text{ turns per second}$$

We calculated the equivalent:

$$f = 4.9 \text{ turns per minute}$$

"This has been an historic day," the king said. "For the first time we have put the new trigonometric functions to practical use. We'll call this new bridge the Trigonometry Memorial Bridge. To commemorate that fact, we will put a sign by the side of the road."

Trigonometeris blushed. "Then we should put a cosine on the other side of the road so as to give equal honor to the two functions that helped us get this far."

Note to CHAPTER 4

- The magnitude of a force can be measured by a unit called the *newton*. One newton equals one kilogram-meter per second squared. In other words, a force of 1 newton will acclerate a mass of 1 kilogram at the rate of 1 meter per second per second. For example, an object with a mass (m) of 20 kilograms will be pulled on by a force of gravity mg. Since $g = 9.8$ meters per second squared, the force will be 196 kilogram-meters per second squared = 196 newtons.

For Exercises 1 to 7, calculate the east-west component and the north-south component of velocity for the velocity vectors.

1. 10 miles per hour 15° north of east

2. 34 miles per hour 30° south of east

3. 5 miles per hour 12.4° north of west

4. 1 mile per hour 87° north of east

5. 60 miles per hour 34° south of west

6. 200 miles per hour 17° north of west

7. 80 miles per hour northwest

Consider an airplane that always flies directly east (relative to the air). However, the wind always blows directly north, which means that the plane's course relative to the ground does not point directly east. The following table gives the plane's airspeed (its speed relative to the air) and the angle that tells how much it is off course. For Exercises 8 to 14, calculate the wind speed.

	Airspeed	Angle		Airspeed	Angle
8.	50	45°	12.	400	4.3°
9.	100	20°	13.	180	5.4°
10.	490	20°	14.	540	8.8°
11.	600	5°			

The following table lists the initial speed (in meters per second) and the angle of launch for several different objects. For Exercises 15 to 19, calculate the distance that each object will travel until it hits the ground.

	Initial speed	Angle of launch
15.	10	25°
16.	40	25°
17.	60	54°
18.	52	48°
19.	100	5°

Suppose that a book is allowed to slide down a frictionless table of length d meters that is tilted at an angle A. Calculate the time for the book to reach the end of the table. Use the formula: time $= \sqrt{2d/(g\sin A)}$ seconds. (Remember $g = 9.8$.)

	d	A		d	A
20.	1.4	10°	23.	1.35	60°
21.	1.8	12°	24.	10	90°
22.	5	20°	25.	10	0°

26. Consider a football player who runs at a speed of 7 yards per second on an open field. How long will it take him to gain 10 yards if he runs straight downfield? What if his course makes a 10° angle with the sidelines? What about these courses: 20° angle; 30° angle; 40° angle; 50° angle; 60° angle.

   If you look at a stick protruding from a lake, it will seem bent. The reason for this is *refraction*. Refraction refers to the bending of light rays when they travel from one medium (such as air) to another (such as water or glass). The amount of bending can be calculated from *Snell's law*. Let A_1 represent the angle

Figure 4-14

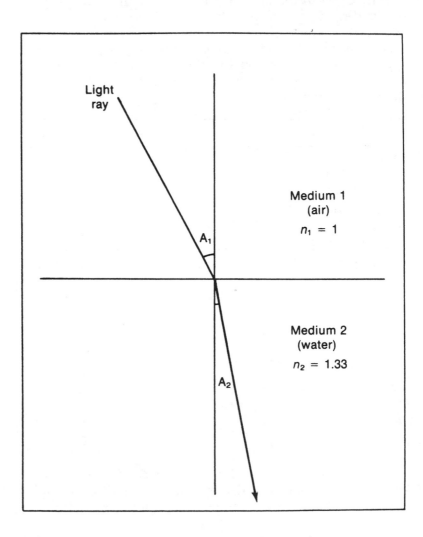

of incidence in medium 1, and let A_2 represent the angle of incidence in medium 2. (See Figure 4-14.) For each medium we need to know the *index* of *refraction*. Let n_1 be the index of refraction in medium 1, and n_2 be the index of refraction in medium 2. Then Snell's law states

$$n_1 \sin A_1 = n_2 \sin A_2$$

The index of refraction for air is $n_1 = 1$; the index of refraction for water is $n_2 = 1.33$. The following table lists values of A_1. For Exercises 27 to 34, calculate $\sin A_2$, and calculate A_2 if possible.

★27. 41.68°   ★28. 0°   ★29. 70.13°   ★30. 16°

★31. 25°   ★32. 30°   ★33. 45°   ★34. 50°

★35. Calculate the value of A_1 if $A_2 = 40°$.

★36. What is the value of A_1 if $A_2 = 48.75°$?

★37. What is the value of A_1 if $A_2 > 48.75°$?

★38. Suppose a light ray passes from air to glass. Suppose $A_1 = 20°$ and $A_2 = 12.34°$. Calculate the index of refraction for the glass.

Radian Measure

That night we camped at the bridge site.
Trigonometeris had enjoyed his first day as Royal
Keeper of the Triangles very much. The next day he
asked Recordis's help in constructing a complete table
of values for the trigonometric functions.
Trigonometeris dearly loved his table of trigonometric
functions, and he wanted a new copy written in an
elegant calligraphic hand. Recordis was glad to know
he was still appreciated. He had been quite jealous the
day before when trigonometry seemed to be getting so
much attention. However, there was no doubt that
Recordis was by far the best person in the kingdom for
writing complicated reports involving long columns of
numbers. He painstakingly constructed a table, starting
at 1°, then 2°, and so on. He wrote all the numbers in
his very best handwriting.

Trigonometeris waited patiently while Recordis
worked all morning. However, suddenly he heard a

tremendous scream. Trigonometeris ran to Recordis and saw him sobbing. "I've ruined it!"

Trigonometeris looked at the parchment.

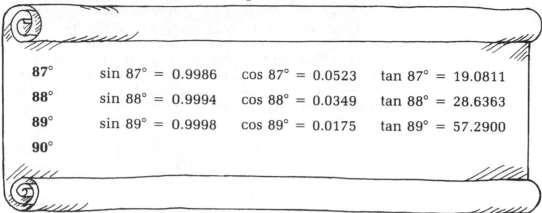

87°	sin 87° = 0.9986	cos 87° = 0.0523	tan 87° = 19.0811
88°	sin 88° = 0.9994	cos 88° = 0.0349	tan 88° = 28.6363
89°	sin 89° = 0.9998	cos 89° = 0.0175	tan 89° = 57.2900
90°			

"What's the matter?" he asked Recordis.

"I should not have written that 90 down!" Recordis sobbed. "I used indelible ink, so I need to start all over again. Everybody knows that there is no such thing as sin 90° or cos 90° or tan 90°."

"Why not?" Trigonometeris asked.

"A right triangle can only have one right angle!" Recordis exploded. "Look at what happens to the Royal Triangles if we set the angle of interest at 90°." (See Figure 5-1.) "The length of the far side becomes the

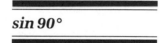

sin 90°

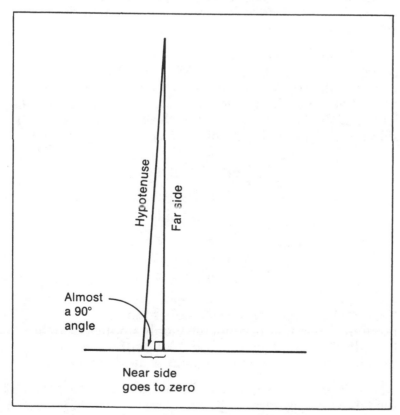

Figure 5-1

same as the length of the hypotenuse, and the length of the near side goes to zero! Then you don't have a triangle any more!"

Recordis continued to sob over this development, but Trigonometeris quickly became excited. "This proves that trigonometric functions are far more versatile than we had previously imagined! When the angle of interest is 90°, then we will say that the length of the near side is zero and the length of the far side is the same as the length of the hypotenuse. Therefore:

$$\sin 90° = \frac{\text{far side}}{\text{hypotenuse}} = 1$$

$$\cos 90° = \frac{\text{near side}}{\text{hypotenuse}} = 0$$

Recordis stared speechlessly at these results. He was quite used to algebra taking totally unexpected turns, but he had thought that trigonometry was a completely straightforward, albeit hopelessly dull, subject.

"But there is no way that you can define a value of $\tan 90°$," Recordis finally said, realizing that Trigonometeris had no way to weasel out of that objection. "Since $\tan A = \sin A/\cos A$, to calculate $\tan 90°$ we would have $\sin 90°/\cos 90° = \frac{1}{0}$. We know that it is totally illegal to have a fraction with a zero on the bottom."

sin 0°

Trigonometeris had to agree with him there. However, he had a new idea. "We can also calculate the values of the trigonometric functions of a zero degree angle," he said. "If the angle of interest is zero, then the length of the far side is zero and the length of the near side is equal to the length of the hypotenuse. Therefore,

$$\sin 0° = \frac{\text{opposite side}}{\text{hypotenuse}} = 0$$

$$\cos 0° = \frac{\text{adjacent side}}{\text{hypotenuse}} = 1$$

$$\tan 0° = \frac{\sin 0}{\cos 0} = 0$$

"I bet some people still adhere to the old-fashioned idea that trigonometry only applies to right triangles," Trigonometeris said. "We will prove that they have never been more mistaken in their lives."

Recordis was glad that he did not have to start the table over again.

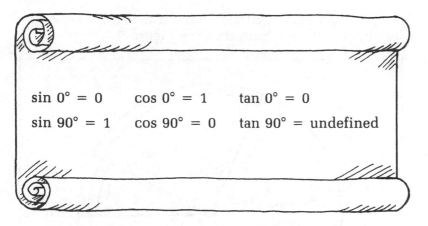

$$\sin 0° = 0 \qquad \cos 0° = 1 \qquad \tan 0° = 0$$
$$\sin 90° = 1 \qquad \cos 90° = 0 \qquad \tan 90° = \text{undefined}$$

Suddenly we were interrupted by an urgent message from Builder. "The gremlin is trying to sabotage the bridge-building process! He is attacking us with a swarm of killer bees!"

By the time we reached Builder we found that he had quickly constructed a solution to the problem. Pal was operating a swiveling ray gun mounted on top of a hill, and the attack of bees was soon under control. Builder explained to us how the device works. "It was no problem to build this," Builder said. "The only tricky part is figuring out how to aim the ray gun. But we have worked out a very good system. The barrel of the gun is 1 meter long. It is designed so that it can rotate about its end. The gun always starts out in the starting position, which points directly to the right. Then I signal Pal to tell him how far the tip of the gun needs to rotate. For example, if I tell Pal to rotate the gun by 1 meter, then he rotates it like this." (See Figure

The Attack of the Killer Bees

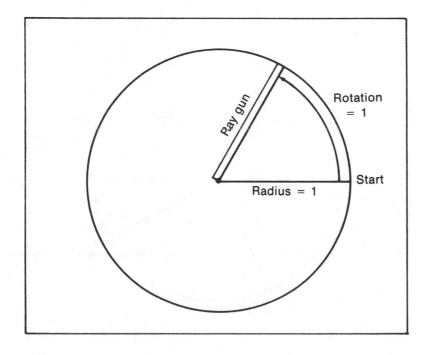

Figure 5-2

5-2.) Builder also illustrated what it meant to rotate the gun by 0.5, 2, and 3 meters. (See Figure 5-3.)

Figure 5-3

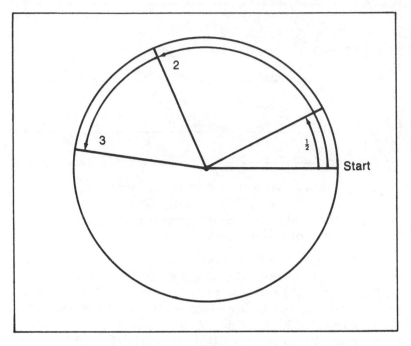

"Normally, we will always be rotating the gun in a counterclockwise direction," Builder explained. "Therefore, whenever I give the rotation amount as a *positive* number, then it means to rotate counterclockwise. However, there may be times when we need to rotate the gun in a clockwise direction. In those cases I give the rotation measurement as a negative number." Builder illustrated some negative rotations. (See Figure 5-4.)

Figure 5-4

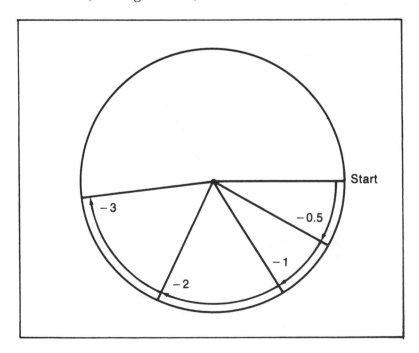

"A totally new concept!" the professor said excitedly. She always became excited when she discovered a totally new concept. "We have never tried to measure rotations before."

"That is *not* a totally new concept," Recordis said. "Measuring rotations is almost exactly the same as measuring angles."

The professor was crestfallen when she realized the obvious similarity between measuring rotations and measuring angles. However, Trigonometeris suddenly became excited about the idea. "This is a totally new way to measure angles!" he said excitedly. "Previously, we have measured angles with degree measure. We have only found a meaning for angles that were greater than 0° and less than 180°. Now we have a new way to measure angles, and we can even define negative angles!"

"It took me a long time before I started believing in negative numbers, so I'm not going to start believing in negative angles!" Recordis fumed.

However, the professor quickly liked the idea. "We will call this new type of measure for angles *radian measure*," she decided. "We are measuring the size of an angle by measuring the distance we must rotate around a circle, and we are expressing that distance as a multiple of the radius of the circle."

The king issued a proclamation.

Radian Measure

Draw a circle of radius r. Draw an angle with the vertex at the center of the circle. (This type of angle will be called a *central angle*.) The two sides of the angle will cut across the circle and form an arc. Let s represent the length of the arc. Then the radian (rad) measure of the angle is

$$\text{Size of angle in radians} = \frac{s}{r}$$

If the angle is formed by rotating counterclockwise, then the angle is a positive angle. If the angle if formed by rotating clockwise, then the angle is a negative angle. (See Figure 5-5.)

Figure 5-5

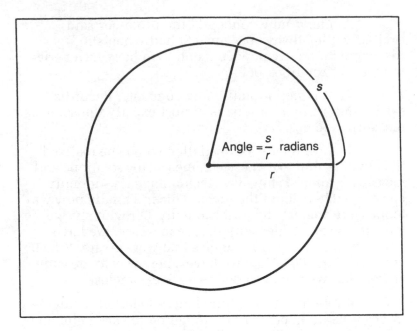

(Note that the circle formed by Builder's ray gun has a radius of 1, so in that case the radian measure of the angle is the same as the length of the arc.)

"We still should find a way to convert angles measured in radian measure into familiar old degree measure," the king added.

"First we must establish exactly how many radians there are in a complete turn," the professor said matter-of-factly.

Builder gave us a clue. "A rotation of 6 radians is a bit less than a complete turn, but a rotation of 7 radians is a bit more than a complete turn. (See Figure 5-6.)

Figure 5-6

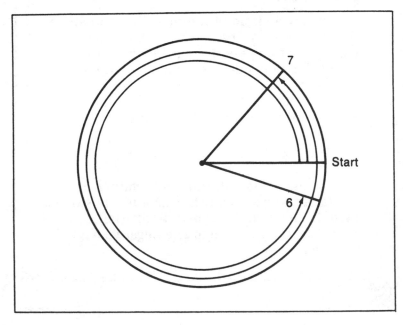

"We need to find a number between 6 and 7," the professor said thoughtfully.

"We were doing some work with circles a long time ago," Recordis said, leafing through his giant record book. "Aha! Here it is. We discovered a special number called pie, symbolized by π, and we found that the circumference of a circle of radius 1 was 2π. (In general, the circumference of a circle of radius r is $2\pi r$.)"

The Special Number π

"The name of that symbol is spelled *pi*, not 'pie,'" the professor corrected. "Pi is the sixteenth letter of the Greek alphabet. Therefore, a complete turn measures 2π radians."

"Then a half-turn must measure π radians," the king said. "And a half-turn is the same as a straight angle, which measures 180°. Therefore,

$$\pi \text{ radians} = 180°$$

We also calculated the radian measure for a quarter-turn (in other words, a right angle) and a three-quarter turn and made a table of results. However, Recordis immediately complained about having to write all the decimal points. We realized that we could simply write radian measures in terms of π—in other words, write π instead of 3.14159, 2π instead of 6.2832, and $\pi/2$ instead of 1.5708. We made a table of results.

Radians	Degrees	
$2\pi = 6.2832$	360	Complete turn
$\pi = 3.1416$	180	Half-turn or straight angle
$\dfrac{\pi}{2} = 1.5708$	90	Quarter-turn or right angle
$\dfrac{\pi}{3} = 1.0472$	60	
$\dfrac{\pi}{4} = 0.7854$	45	
$\dfrac{\pi}{6} = 0.5236$	30	

Converting Radians to Degrees

"We can now state a general formula for converting an angle measured in radians into an angle measured in degrees," Trigonometeris said.

D = angle measured in degrees

R = angle measured in radians

$$D = 180\frac{R}{\pi}$$

"We can also write the reverse formula:

$$R = \frac{\pi D}{180}$$

By using the first formula we found that 1 radian was about equal to 57.296°.

We experimented some more with angles. Pal had fun spinning the ray gun in the direction we told him. We played a game where Recordis told Pal how much to spin while the rest of us hid our eyes. Then we had to figure out the angle that had been formed. One time we looked around and we found that the ray gun was pointed in the starting direction.

"That's easy!" Trigonometeris said. "That's a 0 radian rotation."

"Fooled you!" Recordis said. "It's really a 2π radian rotation."

"How are we supposed to tell the difference between a rotation of 2π (in other words, a complete turn) and a rotation of 0 (in other words, no rotation at all)?" Trigonometeris screamed.

"Those two types of rotations do seem to be effectively identical," the professor said.

"We had better make it illegal to rotate by more than 2π," Trigonometeris said. "For example, a rotation of $(2\pi + \pi/2)$ would be impossible to distinguish from a rotation of $\pi/2$."

Coterminal Angles

"What's wrong with that?" the professor said. "We'll just say that a $(2\pi + \pi/2)$ angle is effectively identical with a $\pi/2$ angle." We decided to use the word *coterminal* to describe the situation where two angles were effectively identical—in other words, their terminal sides were the same. We realized that there were lots of possible angles that are coterminal with a particular angle.

The king declared:

Coterminal Angles

Consider any angle A. This angle is coterminal with the angle $(2\pi + A)$ and the angle $(4\pi + A)$ and the angle $(6\pi + A)$ and the angle $(8\pi + A)$, and so on. In general, the angle $(2n\pi + A)$, where n can be any integer, will be coterminal with A. The values of the trigonometric functions for an angle will be the same for all angles that are coterminal with the original angle. In particular,

$$\sin A = \sin (2n\pi + A)$$

$$\cos A = \cos (2n\pi + A)$$

$$\tan A = \tan (2n\pi + A)$$

for any value of A.

At that moment we were interrupted by the arrival of the Royal Astronomer, who came floating on a small boat down the river. He was carrying a worn knapsack, indicating that he was returning from a long journey. He was pleasantly surprised to see us waiting along the river bank, but as soon as he docked his boat we could see that he was deeply depressed again. "I have just returned from a long journey to distant lands," he explained. "I have been trying to measure the radius of the Earth. I had a brilliant idea for an experiment, but it was totally ruined. I was on South Southsea Island. I was in radio contact with my assistant on North Southsea Island, which is exactly 833 kilometers due north. Before doing the experiment we planned to calibrate our instruments by measuring the position of the sun. That's when we found our instruments were not aligned properly. I measured that the sun was directly overhead (at the point I call the *zenith*), but my assistant found that at that exact same moment the sun was 7.5° away from the zenith. I can't imagine what could have caused an error that large! So I am on my way home to fix the instruments. The whole trip was wasted!" he sobbed.

"We have been having great success with trigonometry," Trigonometeris said. He explained what we had accomplished so far. In an effort to cheer the

The Shifting Sun

astronomer, he offered to convert the 7.5° angle into radian measure.

"Whenever I see a 7.5° angle my mind is filled with painful memories," the astronomer said, but Trigonometeris proceeded anyway.

$$7.5° = \frac{7.5 \times 3.14159}{180} \text{ radians} = 0.1309 \text{ radians}$$

"Here is what that means," the professor explained. 'Suppose you have an arc of length s cut from a circle of radius r by a central angle of 7.5°. Then, $s/r = 0.1309$."

"That's all very interesting, but I'm afraid that this information does me no good," the astronomer said sadly. "It would only help to know this if I was dealing with circles." Suddenly he stopped. All traces of despair vanished in an instant, and he became excited. "That's it!" he exclaimed. "How could I have been so stupid! The Earth is round—and that causes the apparent position of the sun to be different at different locations!" He drew a quick diagram. (See Figure 5-7.)

Figure 5-7

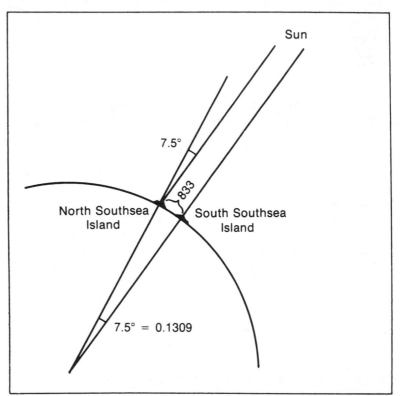

"At South Southsea Island, the sun was directly overhead. But at North Southsea Island, the sun was 7.5° south of the zenith. That means that the lines joining these two islands to the center of the Earth meet to form an angle of 7.5°, or 0.1309 radians."

His eyes suddenly grew even wider. "We can now measure the radius of the Earth!" he gasped. "Let s

be the length of the arc from North Southsea Island to South Southsea Island, and let r be the radius of the Earth. Then, as you have just said:

$$\frac{s}{r} = 0.1309$$

"Therefore,

$$r = \frac{s}{0.1309}$$

"We know $s = 833$. Therefore,

$$r = \frac{833}{0.1309} = 6400 \text{ kilometers (approximately)}$$

The astronomer went home in a state of ecstasy.

"Now we can make general definitions of the trigonometric functions," Trigonometeris said. "Let's start by drawing a line pointing directly to the right."

"That's what we called an x axis," Recordis said, trying to make the situation look more familiar.

"We may as well also add a y axis," the professor said. (See Figure 5-8.)

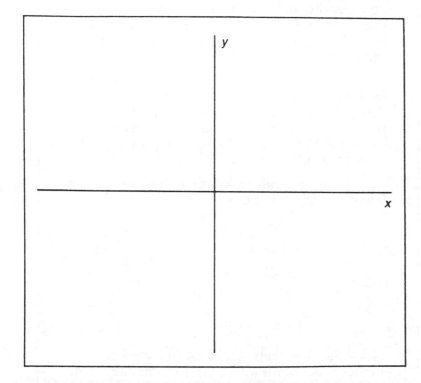

Figure 5-8

Then Trigonometeris suggested how we could give a general definition for the trigonometric functions. There were several quarrels between Trigonometeris and the professor over the exact wording, but here was the result that they finally agreed upon.

Trigonometric Functions

General Definition of Trigonometric Functions

First, draw an xy coordinate system. Then draw an angle in *standard position*. Here is what we mean by standard position: the vertex (point) of the angle is at the origin [the point (0, 0)], and one side of the angle points along the x axis in the positive direction. [We call this side the *initial side*. The other side of the angle is called the *terminal side*. (See Figure 5-9.)

You may measure the size of the angle using either degree measure or radian measure. Here is how to measure the size of the angle using radian measure. Draw a circle with a radius of length 1 centered at the origin. Then the radian measure of the angle is the distance you must travel around the circumference of the circle to get from the initial side (the x axis) to the terminal side. If you travel counterclockwise, then we say that the angle is positive; if you travel clockwise, then the angle is negative. Let's suppose that the angle measures A radians.

Pick any point along the terminal side of the angle. Let's say that the coordinates of this point are (x, y). We will let r represent the distance from the origin to this point. From the Pythagorean theorem we know that $r^2 = x^2 + y^2$.

Now we can make the definitions of the trigonometric functions:

$$\sin A = \frac{y}{r}$$

$$\cos A = \frac{x}{r}$$

$$\tan A = \frac{y}{x}$$

Note that these ratios will be the same no matter what point along the terminal side you pick. (Of course, the value of these ratios will change if you change the angle A.)

Figure 5-9

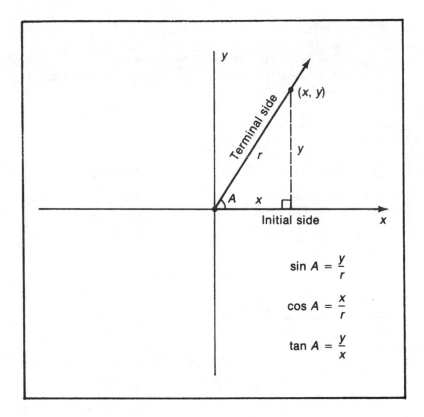

$$\sin A = \frac{y}{r}$$

$$\cos A = \frac{x}{r}$$

$$\tan A = \frac{y}{x}$$

"Now we can begin the systematic study of trigonometric functions," the professor said.

"I just realized something is very wrong with these definitions!" Recordis said. "Sometimes the value of x or y might be negative, so sometimes the value of the trigonometric functions themselves could be negative."

"What's wrong with that?" Trigonometeris asked.

Recordis could not think of a reason why the trigonometric functions could not be negative. We investigated the possibilities. We found that it depends on which *quadrant* the terminal side of the angle is in. There are four possibilities. (See Figure 5-10.)

First quadrant
$\quad$ x, y both positive
$\quad$ Angles from 0 to $\pi/2$ (0 to 90°)
$\quad$ $\sin A$, $\cos A$, and $\tan A$ are all positive

Second quadrant
$\quad$ x negative, y positive
$\quad$ Angles from $\pi/2$ to π (90 to 180°)
$\quad$ $\sin A$ is positive, $\cos A$ is negative, and $\tan A$ is
$\qquad$ negative

Third quadrant
$\quad$ x, y both negative
$\quad$ Angles from π to $3\pi/2$ (180 to 270°)
$\quad$ $\sin A$, $\cos A$ are both negative; $\tan A$ is positive

Fourth quadrant
x positive, y negative
Angles from $3\pi/2$ to 2π (270 to 360°)
$\sin A$ is negative, $\cos A$ is positive, and $\tan A$ is negative

Figure 5-10

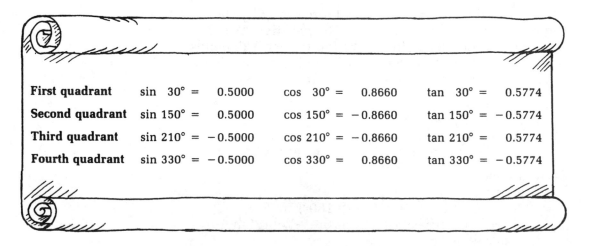

	y	
Quadrant II		Quadrant I
x negative y positive		x, y both positive
cos, tan negative sin positive		sin, cos, tan all positive
	$90° = \dfrac{\pi}{2}$	
$180° = \pi$		$0°$
		x
	$270° = \dfrac{3\pi}{2}$	
sin, cos negative tan positive		sin, tan negative cos positive
Quadrant III		Quadrant IV
x, y both negative		x positive y negative

We found some examples.

First quadrant	$\sin 30° = 0.5000$	$\cos 30° = 0.8660$	$\tan 30° = 0.5774$
Second quadrant	$\sin 150° = 0.5000$	$\cos 150° = -0.8660$	$\tan 150° = -0.5774$
Third quadrant	$\sin 210° = -0.5000$	$\cos 210° = -0.8660$	$\tan 210° = 0.5774$
Fourth quadrant	$\sin 330° = -0.5000$	$\cos 330° = 0.8660$	$\tan 330° = -0.5774$

"The value of the sine function can be negative, but that doesn't mean that it can take on any possible value," Trigonometeris pointed out. We found that neither the sine function nor the cosine function could ever have a value greater than 1 or less than −1, so the king made a decree.

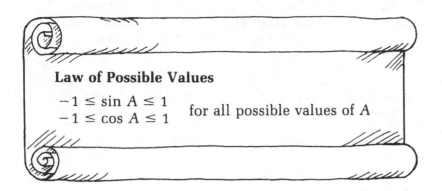

Law of Possible Values

$$-1 \leq \sin A \leq 1$$
$$-1 \leq \cos A \leq 1 \quad \text{for all possible values of } A$$

"It will help to make a list of equations that we know will be true all the time," the king said. "Then, no matter what particular angle we picked, we could know that we could depend on those equations."

"We gave a special name to an equation that is always true," the professor said. "We called it an *identity*. (See the Notes at the end of the chapter.)

We started to make a list of identities. We were able to find simple formulas for the sine, cosine, and tangent of the negative of an angle:

$$\cos(-A) = \cos A$$
$$\sin(-A) = -\sin A$$
$$\tan(-A) = -\tan A$$

(See Exercise 107.)

We also had found that these two relations were true:

$$\cos\left(\frac{\pi}{2} - A\right) = \sin A$$
$$\sin\left(\frac{\pi}{2} - A\right) = \cos A$$

For example, $\cos(\pi/6) = \sin(\pi/2 - \pi/6) = \sin(\pi/3)$.

Cofunctions

"These two equations mean that the sine function and the cosine function are *complementary functions*," the professor said. "In geometry we decided that the *complement* of the angle A was the angle $90° - A$ (which is $\pi/2 - A$ if A is measured in radians). These equations mean that the sine of A is equal to the cosine of the complement of A, and vice versa."

"It almost looks as if the name cosine was set up to mean the complementary function for the sine," Recordis said, looking at Trigonometeris suspiciously.

"Let's make up a new name," the professor said. "Let's say that the cosine function is the *cofunction* for the sine function. And, vice versa, we can say that the sine function is the cofunction for the cosine function."

Cotangent Function

"We should think of a cofunction for the tangent function," Trigonometeris said. "Otherwise it might become lonely."

We decided that we would use the term *cotangent* (abbreviated as ctn or cot) to represent the cofunction for the tangent function. Then we made the definition:

$$\text{ctn}\, A = \tan\left(\frac{\pi}{2} - A\right)$$

To our amazement, we found a very simple expression for the cotangent:

$$\tan A = \frac{y}{x} \qquad \text{ctn}\, A = \frac{x}{y}$$

(See Exercise 108.)

Reciprocal Functions

"They are reciprocals of each other!" the professor said. "It is clear from these equations that $\tan A = 1/\text{ctn}\, A$ and $\text{ctn}\, A = 1/\tan A$."

"But now that the tan function has a reciprocal function, we must find reciprocal functions for the sine and cosine functions," Trigonometeris said. "Otherwise those two functions will become very jealous of the tangent function."

The Secant and Cosecant Functions

Trigonometeris made up a new strange name for the reciprocal of the cosine function. He called it the *secant function* (abbreviated sec):

$$\sec A = \frac{1}{\cos A} \qquad \sec A = \frac{r}{x}$$

It turned out that the reciprocal function for the sine function was also the cofunction for the secant function, so we called it the *cosecant function* (abbreviated csc):

$$\csc A = \frac{1}{\sin A} = \sec\left(\frac{\pi}{2} - A\right) \qquad \csc A = \frac{r}{y}$$

"We have discovered a lot of results today!" Trigonometeris said excitedly as we set up camp for the night.

Summary of Reciprocal Functions and Cofunctions

Function	Reciprocal function	Cofunction
$\sin A$	$\csc A = \dfrac{1}{\sin A}$	$\cos A = \sin\left(\dfrac{\pi}{2} - A\right)$
$\cos A$	$\sec A = \dfrac{1}{\cos A}$	$\sin A = \cos\left(\dfrac{\pi}{2} - A\right)$
$\tan A$	$\operatorname{ctn} A = \dfrac{1}{\tan A}$	$\operatorname{ctn} A = \tan\left(\dfrac{\pi}{2} - A\right)$
$\operatorname{ctn} A$	$\tan A = \dfrac{1}{\operatorname{ctn} A}$	$\tan A = \operatorname{ctn}\left(\dfrac{\pi}{2} - A\right)$
$\sec A$	$\cos A = \dfrac{1}{\sec A}$	$\csc A = \sec\left(\dfrac{\pi}{2} - A\right)$
$\csc A$	$\sin A = \dfrac{1}{\csc A}$	$\sec A = \csc\left(\dfrac{\pi}{2} - A\right)$

"Look how much paper I have used!" Recordis pointed to the piles of papers that contained the results we had discovered that day. "I hope that all this paper won't weigh down the balloon too much. We need to return to Capital City tomorrow."

Notes to
CHAPTER 5

- When the size of an angle is written as a number without a degree symbol, then it is understood that the angle is being measured in radians. Therefore, you can say that the size of an angle is "$\pi/2$" instead of having to say "$\pi/2$ radians." Because the radian measure of an angle is defined as a ratio, it doesn't matter what units you use to measure the distances. The radian measure of an angle is the same whether you use meters or feet or any other units.

- The size of the Earth was calculated by Eratosthenes of Cyrene in 270 B.C. using the method described here. He observed the sun from Alexandria and Syene on the Nile. Christopher Columbus would have had a much better idea about the size of the Earth if he had known about Eratosthenes's calculations.

- Suppose a straight line crosses two parallel lines, as shown in Figure 5-11. The two angles A and B are said to be corresponding angles. In geometry, it can

Figure 5-11

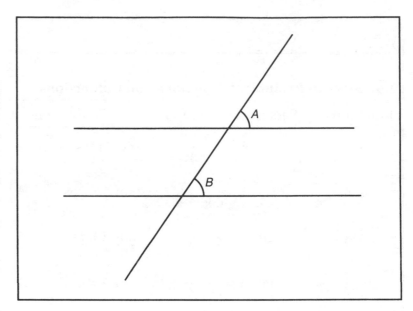

be shown that these two angles are equal. An example of this occurs in Figure 5-7. Because the sun is so far away from the Earth, it can be assumed that the two light rays from the sun are parallel. Then the two angles labeled 7.5° are corresponding angles.

● Some special equations are true for all possible values of the unknowns they contain. Equations of this kind are called *identities*. Here are some examples of identities from algebra.

$$3x = x + x + x$$

$$3(a + b) = 3a + 3b$$

The equation

$$2x = 10$$

is not an identity because there is only one value of x that makes the equation true.

The trigonometric equation

$$\sin A = \cos\left(\frac{\pi}{2} - A\right)$$

is an identity because it is true for any value of A. The equation

$$\sin A = \frac{1}{2}$$

is not an identity because the only solutions are $A = \pi/6$ and $A = 5\pi/6$ and the other angles coterminal with those angles. In the next chapter we will investigate many other trigonometric identities.

● The *domain* of a function is the set of all allowable values for the input number for that function. Because we have now defined values of trigonometric functions for any real number, the domain for each trigonometric function consists of all real numbers. However, there are some exceptions: $\tan(\pi/2)$, $\tan(3\pi/2)$, $\text{ctn}\,0$, $\text{ctn}\,\pi$, $\sec(\pi/2)$, $\sec(3\pi/2)$, $\csc 0$, and $\csc \pi$ are not defined, so these values and their coterminal values are not part of the domain for the functions listed.

The *range* of a function is the set of all possible values of the output number. For the sine and cosine function, the range is from −1 to 1. The ranges of the tangent and cotangent function consist of all real numbers. The ranges of the secant and cosecant functions consist of all real numbers except those between −1 and 1.

For Exercises 1 to 16, convert the angles measured in radians into degrees.

1.	$\pi/3$	7.	$2\pi/5$	13.	1.645
2.	$\pi/6$	8.	1	14.	2.9875
3.	$\pi/4$	9.	2	15.	3.645
4.	$\pi/5$	10.	3	16.	1.987
5.	$\pi/10$	11.	4		
6.	$\pi/12$	12.	5		

For Exercises 17 to 31, convert the angles measured in degrees into angles measured in radians.

17.	30°	23.	1°	29.	1 minute
18.	45°	24.	57°	30.	1 second
19.	270°	25.	58°	31.	5° 12 minutes
20.	100°	26.	60°		16 seconds
21.	216°	27.	80°		
22.	4.5°	28.	85°		

Identify the angle between 0 and 2π that is coterminal with each of the angles in Exercises 32 to 40.

32.	16π	35.	100π	38.	12.45π
33.	18π	36.	-0.5π	39.	16.45π
34.	23.6π	37.	-1.5π	40.	14.5π

For Exercises 41 to 50, make a table listing A (measured in radians) and $\sin A$ for the angles listed.

41.	4°	45.	2°	49.	0.2°
42.	3.5°	46.	1.5°	50.	0.1°
43.	3°	47.	1°		
44.	2.5°	48.	0.5°		

51. Can you suggest an approximation for $\sin A$ as A becomes small?

For Exercises 52 to 69, calculate $\sin A$, $\cos A$, and $\tan A$ for these angles (in radians). Also calculate the measure of these angles in degrees. (Do not use a calculator for Exercises 52 to 62.) For Exercises 63 to 69, be sure your calculator is set in radian mode.

52.	π	58.	$5\pi/6$	64.	2
53.	$3\pi/2$	59.	$7\pi/6$	65.	3
54.	$3\pi/4$	60.	$4\pi/3$	66.	4
55.	$5\pi/4$	61.	$5\pi/3$	67.	5
56.	$7\pi/4$	62.	$11\pi/6$	68.	6
57.	$2\pi/3$	63.	1	69.	7

In Exercises 70 to 74, you are given values for $\sin A$ and $\cos A$. Determine the value of A.

70. $\sin A = 1/\sqrt{2}$; $\cos A = 1/\sqrt{2}$

71. $\sin A = -\frac{1}{2}$; $\cos A = \sqrt{3}/2$

72. $\sin A = -1$; $\cos A = 0$

73. $\sin A = \sqrt{3}/2$; $\cos A = -\frac{1}{2}$

74. $\sin A = \frac{1}{2}$; $\cos A = \sqrt{3}/2$

In Exercises 75 to 80, you are given values for $\sin A$ and $\tan A$. Determine the value of A.

75. $\sin A = -1/\sqrt{2}$; $\tan A = 1$

76. $\sin A = 1$; $\tan A$ undefined

77. $\sin A = -\frac{1}{2}$; $\tan A = -1/\sqrt{3}$

78. $\sin A = \sqrt{3}/2$; $\tan A = -\sqrt{3}$

79. $\sin A = 1/\sqrt{2}$; $\tan A = -1$

80. $\sin A = -\frac{1}{2}$; $\tan A = 1/\sqrt{3}$

Figure 5-12

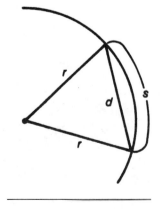

★ 81. Suppose you know the length s of an arc on a circle of radius r. Calculate the length of the associated chord—that is, find the distance between the two end points of the arc. (See Figure 5-12.)

★ 82. Consider a runner running around a perfectly circular track of radius 25 meters. Suppose you measure the central angle between the starting point and the runner's current position, and you find that this angle is increasing at a constant speed of 0.2256 radians per second. (The rate of increase of this angle is called the *angular velocity*.) How fast is the runner running?

★ 83. Derive a general formula that relates r (the radius of the track), v (the runner's speed), and ω (the angular velocity).

★ 84. Let's suppose that the planets orbit the sun at constant speeds around perfectly circular orbits. (In reality the planet's orbits are ellipses, but they are close to being circles.) Calculate the orbital velocity (in kilometers per day) and the angular velocity (in radians per day) for each planet, given the information in the following table.

Planet	Radius of orbit (million kilometers)	Period of orbit (days)
Mercury	58	88
Venus	108	225
Earth	150	365
Mars	228	687
Jupiter	778	4,333
Saturn	1,427	10,759

In Exercises 85 to 88, you are given the distance in kilometers between two points that are on the same north/south line and the angular difference between the sun's position seen from these two positions. Calculate the radius of the planet that each pair of points is located on.

85. 10°; 1113   87. 2°; 2492

86. 30°; 1275   88. 34°; 3590

Write formulas for the exact values for the trigonometric functions in Exercises 89 to 91.

89. sec 30°; csc 30°; ctn 30°

90. sec 45°; csc 45°; ctn 45°

91. sec 60°; csc 60°; ctn 60°

Calculate values for the trigonometric functions in Exercises 92 to 100. Use a calculator. You will need to use the reciprocal functions since your calculator most likely does not have keys for sec, csc, and ctn. For example, to find sec 20°, use the formula sec 20° = 1/cos 20°.

92. sec 20°   95. sec 11.4°   98. sec 150°

93. ctn 35°   96. ctn 20.8°   99. csc 95°

94. csc 75°   97. csc 63°   100. ctn 170°

In Exercises 101 to 106, calculate an angle between 0 and -2π that is coterminal with each of the given angles.

101. $3\pi/2$ 103. $21\pi/11$ 105. $9\pi/10$

102. $5\pi/6$ 104. $17\pi/11$ 106. $12\pi/14$

107. Show that $\sin(-A) = -\sin A$. Show that $\cos(-A) = \cos A$.

108. Show that $\operatorname{ctn} A = 1/\tan A$.

109. Write a program that prints a table of the sine, cosine, and tangent functions for angles from 0° to 90°. Most programming languages, such as BASIC or Pascal, will automatically calculate these values for you.

6

Trigonometric Identities

We received a letter from Builder two days after our return. Construction on the bridge was going well, but he did have some complaints. "There are still some problems that take too long to solve," he wrote. "For example, often I will know the value of $\sin A$ for a particular angle A, but I need to know the value of $\cos A$. I wish there were a quick formula that would tell me the value of $\cos A$ in that case, so I wouldn't have to look it up in the table again."

"No way!" Recordis exclaimed. "Sines and cosines are fundamentally different entities—there is no way to find a connection between them."

"We could write down the defining relations and see if something hits us," the professor said encouragingly.

$$\sin A = \frac{y}{r} \quad \cos A = \frac{x}{r}$$

"All we need is to find a relation between x, y, and r."

"There is no relation between x, y, and r!" Recordis exclaimed. "Except, of course, for the Pythagorean theorem," he reluctantly added. (Recordis passionately disliked square root signs, so he was reluctant to use the Pythagorean theorem because the results often involved square root signs.)

"We will use the Pythagorean theorem!" the professor exclaimed.

$$x^2 + y^2 = r^2$$

"Divide both sides of that equation by r^2," Trigonometeris suggested.

$$\frac{x^2}{r^2} + \frac{y^2}{r^2} = \frac{r^2}{r^2}$$

"We know $r^2/r^2 = 1$," the king volunteered.

$$\frac{x^2}{r^2} + \frac{y^2}{r^2} = 1$$

Pythagorean Identities

"Now, use the definition of $\sin A$ and $\cos A$!" Trigonometeris said eagerly.

$$\frac{y^2}{r^2} = \sin^2 A \qquad \frac{x^2}{r^2} = \cos^2 A$$

"Therefore,

$$\sin^2 A + \cos^2 A = 1$$

"That is another identity—it will be true for any possible value of A."

"I should have known!" Recordis cried. "The old finding-a-relation-between-the-sine-and-the-cosine-by-using-the-Pythagorean-theorem trick!"

We wrote down two obvious equations that followed directly from this first equation.

$$\sin^2 A = 1 - \cos^2 A \qquad \cos^2 A = 1 - \sin^2 A$$

"We can also say

$$\sin A = \sqrt{1 - \cos^2 A}$$

and

$$\cos A = \sqrt{1 - \sin^2 A}$$

but we must be careful when using these formulas because the values of $\cos A$ and $\sin A$ are not always positive," the king said.

"This works in theory," Recordis cautioned, "but we should try an example to make sure it works in practice."

We considered an angle $A = 35°$. We found $\sin A$ = 0.5736 and $\cos A$ = 0.8192. Then we calculated

$$\sin^2 A + \cos^2 A = 0.329 + 0.671$$
$$= 1.000$$

"See! It does work!" the professor said happily.

"We will also be able to find corresponding equations relating the other trigonometric functions," Trigonometeris said. Starting from the equation $x^2 + y^2 = r^2$ and dividing both sides by x^2, we found

$$\tan^2 A + 1 = \sec^2 A$$

Dividing both sides by y^2 we found

$$\text{ctn}^2 A + 1 = \csc^2 A$$

"These identities will be very useful," Trigonometeris said, "and the best part is that we know that they will always be true, no matter what angles we use. I think identities are far more dependable than regular equations."

"Builder has another problem," Recordis said. He continued reading from Builder's letter. "There are often times when I need to stack one triangle on top of another triangle. In those circumstances it would be nice to have a formula for the sine of the sum of two angles; in other words, can you tell me how to calculate $\sin(A + B)$ if I know the values of the trigonometric functions for angle A and angle B?"

"Let's make up a trigonometric addition rule," Recordis said. "I suggest that we make up this rule:

$$\sin(A + B) = \sin(A) + \sin(B)?$$

"I guarantee you that this rule will make life much simpler in the long run."

"But it does not work," the king said. "We know $\sin(30° + 60°) = \sin(90°) = 1$

"But

$$\sin(30°) + \sin(60°) = 0.5000 + 0.8660$$
$$= 1.366$$

"Therefore, $\sin(30° + 60°)$ does not equal $\sin(30°) + \sin(60°)$."

"Besides, we can't just make up a rule like that," the professor said.

"I thought we were making most of this up anyway," Recordis objected. Nevertheless, he agreed to help with the search for a general formula for $\sin(A + B)$.

"In cases such as this, the first thing to do is draw a picture," Trigonometeris suggested.

Figure 6-1

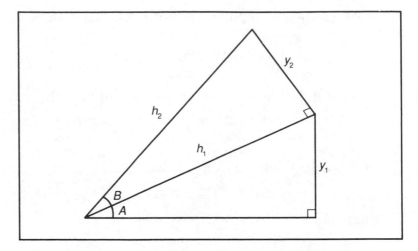

We drew two right triangles stacked on top of each other, one containing the angle A and the other containing the angle B. (See Figure 6-1.)

"I can see from the picture that these two equations are true," Recordis said.

$$y_1 = h_1 \sin A$$

$$y_2 = h_2 \sin B$$

"But we can't find an expression for $\sin(A + B)$ using these triangles. If only we had a right triangle involving the angle $A + B$, like this," he moaned. He drew an addition to our original diagram. (See Figure 6-2.)

Figure 6-2

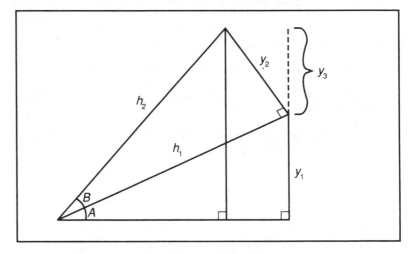

"That's exactly what we need to do!" the professor said. "What can see that:

$$\sin(A + B) = \frac{y_1 + y_3}{h_2}$$

"We don't know an expression for y_3," Recordis pointed out. However, Trigonometeris had an idea. He drew another line segment on the diagram and labeled some of the angles. (See Figure 6-3.)

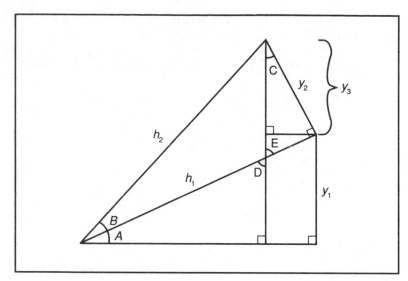

Figure 6-3

"We can see that $y_3 = y_2 \cos C$," he said. To our surprise, we found that angle C was the same size as angle A. To show this, we had to look in an old book on geometry. We could tell that angle D was equal to 90 $- A$, since they were two angles in a right triangle. Then we could see that angle E was equal to angle D, since they are two opposite angles formed by the intersection of two straight lines (these are called vertical angles). Angle C equals $90 - E$, since they are two angles in a right triangle. Putting all these together, we found $A = C$. Therefore:

$$y_3 = y_2 \cos A$$

We had already found $y_1 = h_1 \sin A$, and Recordis recognized that we could substitute these two expressions into our formula for $\sin(A + B)$:

$$\sin(A + B) = \frac{h_1 \sin A + y_2 \cos A}{h_2}$$

Trigonometeris told us two more relations:

$$h_1 = h_2 \cos B$$

$$y_2 = h_2 \sin B$$

Then we used the substitution principle again:

$$\sin(A + B) = \frac{h_2 \cos B \sin A + h_2 \sin B \cos A}{h_2}$$

"We can cancel out all the h_2 terms!" Recordis said excitedly.

Addition Rules

$$\sin(A + B) = \cos B \sin A + \sin B \cos A$$

We decided to change the order of the first term:

$$\sin(A + B) = \sin A \cos B + \sin B \cos A$$

"What an elegant formula!" Trigonometeris said. "The sines and the cosines work together so well."

"We should test some examples to make sure it really works," the professor cautioned.

$$\sin(A + 0) = \sin A \cos 0 + \sin 0 \cos A = \sin A$$

$$\sin\left(\frac{\pi}{3} + \frac{\pi}{6}\right) = \sin\frac{\pi}{3}\cos\frac{\pi}{6} + \sin\frac{\pi}{6}\cos\frac{\pi}{3}$$

$$= \frac{\sqrt{3}}{2} \times \frac{\sqrt{3}}{2} + \frac{1}{2} \times \frac{1}{2}$$

$$= \frac{3}{4} + \frac{1}{4} = 1$$

$$= \sin\frac{\pi}{2}$$

$$\sin(10° + 20°) = \sin 10° \cos 20° + \sin 20° \cos 10°$$

$$= (0.1736)(0.9397) + (0.3420)(0.9848)$$

$$= 0.16313 + 0.33680$$

$$= 0.49993$$

"That's close enough to 0.5 for my purposes," Recordis said, "And we know that $\sin 30° = 0.5$. However, I bet the formula doesn't work if $A + B$ is greater than 180°. For example, $\sin(150° + 40°)$ should be $\sin 190°$, which should be the same as $\sin(180° - 190°) = \sin(-10°) = -0.1736$."

The sweat built up on Trigonometeris's brow while we tried the formula

$$\sin 150° \cos 40° + \sin 40° \cos 150°$$

$$= (0.5)(0.7660) + (0.6428)(-0.8660)$$

$$= 0.383 - 0.5567$$

$$= -0.1737$$

"It does work!" Trigonometeris breathed a sigh of relief.

Since $\sin(-q) = -\sin q$ and $\cos(-q) = \cos q$ for any q, we found a formula for the sine of the difference of two angles:

$$\sin(A - B) = \sin[A + (-B)]$$

$$= \sin A \cos(-B) + \sin(-B)\cos A$$

$$= \sin A \cos B - \sin B \cos A$$

"We can find another elegant formula for $\cos(A + B)$," Trigonometeris said. "We just have to use the fact of nature that $\cos(q) = \sin(90° - q)$ for any value of q."

$$\cos(A + B) = \sin[90° - (A + B)]$$

$$= \sin[(90° - A) - B]$$

$$= \sin(90° - A)\cos B - \sin B \cos(90° - A)$$

$$= \cos A \cos B - \sin B \sin A$$

$$= \cos A \cos B - \sin A \sin B$$

"Now we really have momentum!" Trigonometeris said. He suggested that we look for a formula for $\tan(A + B)$:

$$\tan(A + B) = \frac{\sin(A + B)}{\cos(A + B)}$$

$$= \frac{\sin A \cos B + \sin B \cos A}{\cos A \cos B - \sin A \sin B}$$

Recordis thought this formula was as simple as possible, but Trigonometeris became obsessed with the idea of finding a simpler form for it. He tried several ideas that didn't work, but he finally suggested multiplying both the top and the bottom by $1/(\cos A \cos B)$:

$$\frac{\dfrac{\sin A \cos B}{\cos A \cos B} + \dfrac{\sin B \cos A}{\cos B \cos A}}{\dfrac{\cos A \cos B}{\cos A \cos B} - \dfrac{\sin A \sin B}{\cos A \cos B}}$$

The result was

$$\tan(A + B) = \frac{\tan A + \tan B}{1 - \tan A \tan B}$$

Recordis's eyes were becoming bleary by now, but he thought of a simple idea before anyone else did. "Suppose that $A = B$. Then we know that $\sin(A + B) = \sin(2A)$, so therefore

$$\sin(2A) = \sin A \cos A + \sin A \cos A$$

$$= 2 \sin A \cos A$$

"We'll call that a double-angle formula, since it tells us how to calculate the sine of an angle after you double it," the professor said. We found double-angle formulas for the cosine and tangent functions:

$$\cos(2A) = \cos^2 A - \sin^2 A$$

$$= 1 - 2 \sin^2 A$$

$$= 2 \cos^2 A - 1$$

(Note that we used the identity $\sin^2 A + \cos^2 A = 1$ to write this formula in three different forms. There was a big argument about which form would be the simplest form to use, so we decided we would use all three forms.)

$$\tan(2A) = \frac{2 \tan A}{1 - \tan^2 A}$$

"I see something else," the professor realized. She had become jealous when the others seemed to find

Double-Angle Rules

ideas before she had. "Suppose we known $\sin^2 A$, but we would like a simpler formula with no exponent. We can see from the formula for $\cos 2A$ that

$$\sin^2 A = \frac{1}{2}(1 - \cos 2A)$$

"Also

$$\cos^2 A = \frac{1}{2}(1 + \cos 2A)$$

Trigonometeris gathered all our results together in one list so we could send them to Builder. He grouped them under different headings that described where the identities had come from. (We discovered a few more identities that are included at the end of the list. See the exercises for derivations of these.)

Trigonometric Identities

These equations are true for every possible value of the angles A and B.

Reciprocal functions

$$\sin A = \frac{1}{\csc A} \qquad \csc A = \frac{1}{\sin A}$$

$$\cos A = \frac{1}{\sec A} \qquad \sec A = \frac{1}{\cos A}$$

$$\tan A = \frac{1}{\operatorname{ctn} A} \qquad \operatorname{ctn} A = \frac{1}{\tan A}$$

Cofunctions (radian form)

$$\sin A = \cos\left(\frac{\pi}{2} - A\right) \qquad \cos A = \sin\left(\frac{\pi}{2} - A\right)$$

$$\tan A = \operatorname{ctn}\left(\frac{\pi}{2} - A\right) \qquad \operatorname{ctn} A = \tan\left(\frac{\pi}{2} - A\right)$$

$$\sec A = \csc\left(\frac{\pi}{2} - A\right) \qquad \csc A = \sec\left(\frac{\pi}{2} - A\right)$$

Negative angle relations

$$\sin(-A) = -\sin A$$
$$\cos(-A) = \cos A$$
$$\tan(-A) = -\tan A$$

Quotient relations

$$\tan A = \frac{\sin A}{\cos A}$$

$$\operatorname{ctn} A = \frac{\cos A}{\sin A}$$

Supplementary angle relations

The angles A and B are supplementary angles if $A + B = \pi$.

$$\sin(\pi - A) = \sin A$$
$$\cos(\pi - A) = -\cos A$$
$$\tan(\pi - A) = -\tan A$$

Pythagorean identities

$$\sin^2 A + \cos^2 A = 1$$
$$\tan^2 A + 1 = \sec^2 A$$
$$\operatorname{ctn}^2 A + 1 = \csc^2 A$$

Functions of the sum of two angles

$$\sin(A + B) = \sin A \cos B + \sin B \cos A$$
$$\cos(A + B) = \cos A \cos B - \sin A \sin B$$
$$\tan(A + B) = \frac{\tan A + \tan B}{1 - \tan A \tan B}$$

Functions of the difference of two angles

$$\sin(A - B) = \sin A \cos B - \sin B \cos A$$
$$\cos(A - B) = \cos A \cos B + \sin A \sin B$$
$$\tan(A - B) = \frac{\tan A - \tan B}{1 + \tan A \tan B}$$

Double-angle formulas

$$\sin(2A) = 2 \sin A \cos A$$
$$\cos(2A) = \cos^2 A - \sin^2 A$$
$$= 1 - 2 \sin^2 A$$
$$= 2 \cos^2 A - 1$$
$$\tan(2A) = \frac{2 \tan A}{1 - \tan^2 A}$$

Squared formulas

$$\sin^2 A = \frac{1}{2}(1 - \cos 2A)$$
$$\cos^2 A = \frac{1}{2}(1 + \cos 2A)$$

Half-angle formulas

$$\sin\left(\frac{A}{2}\right) = \pm\sqrt{\frac{1 - \cos A}{2}}$$
$$\cos\left(\frac{A}{2}\right) = \pm\sqrt{\frac{1 + \cos A}{2}}$$

$$\tan\left(\frac{A}{2}\right) = \pm\sqrt{\frac{1+\cos A}{1+\cos A}}$$

Product formulas

$$\sin A \cos B = \frac{1}{2}[\sin(A+B) + \sin(A-B)]$$

$$\cos A \sin B = \frac{1}{2}[\sin(A+B) - \sin(A-B)]$$

$$\cos A \cos B = \frac{1}{2}[\cos(A+B) + \cos(A-B)]$$

$$\sin A \sin B = -\frac{1}{2}[\cos(A+B) - \cos(A-B)]$$

Sum formulas

$$\sin A + \sin B = 2\sin\frac{A+B}{2}\cos\frac{A-B}{2}$$

$$\cos A + \cos B = 2\cos\frac{A+B}{2}\cos\frac{A-B}{2}$$

Difference formulas

$$\sin A - \sin B = 2\cos\frac{A+B}{2}\sin\frac{A-B}{2}$$

$$\cos A - \cos B = -2\sin\frac{A+B}{2}\sin\frac{A-B}{2}$$

Note to CHAPTER 6

- It is important to note that these identities are only true provided that all the arguments for the trigonometric functions have permissible values. For example, any identity involving the tangent function will be unusable if one of the angles has the value 90°.

Exercises

1. Suppose you know the value of $\sin A$ for an angle in the first quadrant. Write equations for $\cos A$, $\tan A$, $\operatorname{ctn} A$, $\sec A$, and $\csc A$ in terms of $\sin A$.

If $\sin A = \frac{3}{5}$ and $\cos A$ is negative, find the value of the trigonometric expressions in Exercises 2 to 6.

2. $\cos A$

2. $\sin 2A$

4. $\tan 2A$

5. $\cos 2A$

6. $\sin(A/2)$

7. If $\tan A = \frac{3}{4}$ and $\cos B = -\frac{5}{13}$, where A and B are both third-quadrant angles, find $\sin(A + B)$.

8. If A is a first-quadrant angle with $\sin A = \frac{12}{13}$ and B is a second-quadrant angle with $\cos B = -\frac{4}{5}$, find $\cos(A + B)$.

Find exact formulas for the trigonometric expressions in Exercises 9 to 11.

9. $\sin 15°$

10. $\sin 75°$

11. $\sin 7.5°$

12. Find an exact formula for $\sin 195°$ by using $\sin 195 = \sin(150° + 45°)$.

13. Find an exact formula for $\sin 75° + \sin 15°$.

Prove the identities in Exercises 14 to 21 using the trigonometric addition formulas.

14. $\sin(-a) = -\sin a$

15. $\cos(-a) = \cos a$

16. $\tan(-a) = -\tan a$

17. $\cos(\pi/2 - a) = \sin a$

18. $\sin(\pi/2 - a) = \cos a$

19. $\tan(\pi/2 - a) = 1/\tan a$

20. $\sec(A + B) = (\sec A \sec B)/(1 - \tan A \tan B)$

21. $\csc(A + B) = (\csc A \csc B)/(\text{ctn}\, A + \text{ctn}\, B)$

In general, to prove a trigonometric identity to be true, you must manipulate one side of the identity until it becomes the same as the other side. (Note that this procedure is different from the procedure you use to solve a conditional equation; there you perform operations on both sides of the equation at the same time.) For example, suppose we need to prove the identity

$$\sin^2 A - \frac{1}{2}(1 - \cos 2A)$$

In most cases the best strategy is to start with the most complicated side and try to transform it to match the simpler side. Here's how to do our example:

$$\frac{1}{2}(1 - \cos 2A) = \frac{1}{2}[1 - (\cos^2 A - \sin^2 A)]$$

$$= \frac{1}{2}[1 - \cos^2 A + \sin^2 A]$$

$$= \frac{1}{2}[\sin^2 A + \sin^2 A]$$

$$= \sin^2 A$$

Prove the identities in Exercises 22 to 43.

22. $\cos^2 A = \frac{1}{2}(1 + \cos 2A)$

★23. $\sin(A/2) = \sqrt{(1 - \cos A)/2}$

★24. $(\sin A)(\cos B) = \frac{1}{2}[\sin(A + B) + \sin(A - B)]$

★25. $\sin A + \sin B = 2\sin[(A + B)/2]\cos[(A - B)/2]$

★26. $\cos A - \cos B = -2\sin[(A + B)/2]\sin[(A - B)/2]$

27. $\sec^2 A + \csc^2 A = (\sec^2 A)(\csc^2 A)$

28. $\sin(A + B + C) = \sin A \cos B \cos C + \cos A \sin B \cos C \\ \qquad\qquad\qquad + \cos A \cos B \sin C - \sin A \sin B \sin C$

29. $\cos(A + B + C) = \cos A \cos B \cos C - \sin A \sin B \cos C \\ \qquad\qquad\qquad - \sin A \cos B \sin C - \cos A \sin B \sin C$

30. $\tan(2A) = (2\tan A)/(1 - \tan^2 A)$

★31. $\sin(4A) = \cos A(4\sin A - 8\sin^3 A)$

★32. $\sin(5A) = 5\sin A - 20\sin^3 A + 16\sin^5 A$

★33. $\cos(3A) = 4\cos^3 A - 3\cos A$

★34. $\cos(4A) = 8\cos^4 A - 8\cos^2 A + 1$

★35. $\sin^3 A = \frac{1}{4}[-\sin(3A) + 3\sin A]$

★36. $\sqrt{(1 + \sin A)/(1 - \sin A)} = \sec A + \tan A$

37. $(\sin A + \cos A)^2 = \sin(2A) + 1$

38. $\sec^4 A - \sec^2 A = \tan^4 A + \tan^2 A$

39. $[\sin(2A)]/(\sin A) - [\cos(2A)]/(\cos A) = \sec A$

★40. $[\sin(3A) - \sin A]/(\cos^2 A - \sin^2 A) = 2\sin A$ (see Exercise 44.)

41. $\operatorname{ctn} B \sec B = \csc B$

42. $\cos A + \sin A \tan A = \sec A$

★43. $(\sin A + \tan A)/(1 + \sec A) = \sin A$

★44. Derive a formula for $\sin(3A)$ in terms of $\sin A$ and $\sin^3 A$.

□45. Write a program that reads in the value of $\sin A$, and reads in the quadrant containing A, and then calculates $\cos A$, $\tan A$, $\operatorname{ctn} A$, $\sec A$, and $\csc A$, using the identities derived in the chapter.

Law of Cosines and Law of Sines

The next day we received an urgent letter from the panic-stricken Royal Construction Engineer. "Help!" Builder wrote. "The gremlin threatened us this morning! He pointed out that triangles are still very mysterious. If we know some parts of a triangle, we can't always calculate the other parts."

"We can use the trigonometric functions!" Trigonometeris exclaimed.

"I know Trigonometeris will say that we can use the trigonometric functions, but it is not that simple." Recordis continued to read Builder's letter. "We can easily solve for the unknown parts of any right triangle. However, just yesterday I was forced to deal with a triangle that I know contains two sides, each 10 meters long, and the angle between these two sides is a 100° angle, but I need to know the length of the third side."

"We must put a stop to the gremlin's threats!" the king cried.

"We should be able to find a general relationship that works for all triangles," Trigonometeris said. "I am sure that the trigonometric functions will come to our rescue once more." Trigonometeris drew an arbitrary triangle on the board. He used a, b, and c to represent the lengths of the three sides, and he used A to represent the angle opposite side a, the letter B to represent the angle opposite side b, and C to represent the angle opposite side c.

Trigonometeris stared at the triangle all morning and into the afternoon. However, he was unable to come up with any ideas for a general formula that would relate a, b, c, A, B, and C.

"I told you things would be much simpler if you had a right triangle," Recordis told him. "I know how to break a nonright triangle into two right triangles. All we need to do is draw the *altitude* of the triangle." (An altitude of a triangle is a line segment perpendicular to one side of the triangle that connects that side to the opposite vertex.) In this case we used the letter h to represent the length of the altitude. (See Figure 7-1.)

Figure 7-1

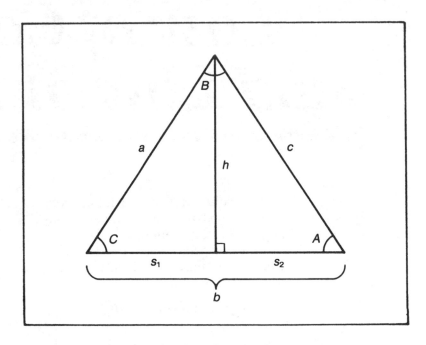

"We can use the Pythagorean theorem for each of those little triangles," the king suggested.

$$h^2 + s_2^2 = c^2$$

"But Builder wants the answer for c expressed in terms of a, b, and C, not h and s_2," Trigonometeris protested.

"We know these equations are true," the professor suggested:

$$h = a \sin C \qquad s_2 = b - s_1$$

$$s_1 = a \cos C$$

"Therefore, $\qquad s_2 = b - a \cos C$

"We can now use the substitution principle from algebra," Recordis said. "The substitution principle says that if two quantities are equal, we have the right to substitute one quantity for the other in any equation. In our case we want to substitute the expressions we have found for h and s_2 into the equation

$$h^2 + s_2^2 = c^2$$

The result was

$$c^2 = (a \sin C)^2 + (b - a \cos C)^2$$

$$= a^2 \sin^2 C + b^2 - 2ab \cos C + a^2 \cos^2 C$$

We rewrote that equation as

$$c^2 = a^2(\sin^2 C + \cos^2 C) + b^2 - 2ab \cos C$$

"We know $\sin^2 C + \cos^2 C = 1$, for any value of C!" Trigonometeris exclaimed, elated that the identity we had discovered just the day before had already turned out to have a practical application. Therefore,

$$c^2 = a^2 + b^2 - 2ab \cos C$$

"That formula is much simpler than I thought it could possibly be," Recordis admitted. "And it is just the formula Builder needs to solve his current problem." (See Figure 7-2.)

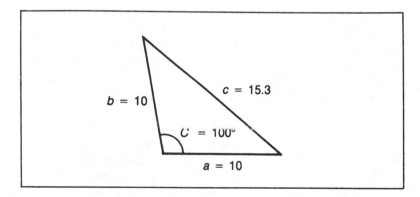

Figure 7-2

We inserted the values $a = 10$, $b = 10$, and $C = 100°$ into the formula, and came up with the result

$$c^2 = 10^2 + 10^2 - 2 \times 10 \times 10 \cos 100°$$

$$= 234.73$$

$$c = 15.3$$

"Every triangle in the world will have no choice but to obey this law," Trigonometeris said.

"We must think of a good name for it," Recordis said.

We decided to call it the law of cosines since it contained a cosine.

Law of Cosines

> The law of cosines is useful when you know two sides of a triangle and the angle between those two sides. Let a and b represent the lengths of the two sides, and let C represent the angle between these two sides. Then the third side (c) can be found from the formula
>
> $$c^2 = a^2 + b^2 - 2ab \cos C$$

"Hold everything!" Recordis suddenly panicked. "You said this rule holds for all triangles. But we know it cannot hold for right triangles, because then the Pythagorean theorem holds:

$$c^2 = a^2 = b^2$$

For a few awful moments we were filled with dread. However, the professor saw a way out of the dilemma. "Look at what happens if $C = 90°$—in other words, if the triangle is a right triangle. Then, $\cos C = 0$, and then the law of cosines becomes the same as the regular Pythagorean theorem,

$$c^2 = a^2 + b^2$$

"We were lucky that time," Recordis breathed a sigh of relief. "It would have been terrible if we discovered a new rule that contradicted something we had done before, especially something as vitally important as the Pythagorean theorem."

The king noticed another interesting feature. If C is less than $90°$, then $\cos C$ is positive and c^2 will be less than $a^2 + b^2$. On the other hand, if C is greater than $90°$, then $\cos C$ is negative and c^2 will be greater than $a^2 + b^2$.

As we continued to read Builder's letter we found another problem. "I have another triangle for which the three angles must be 80°, 60°, and 40°. I know that the side opposite the 80° angle must be 10 meters long, but I need to know the lengths of the other two sides."

We found that we could not use the law of cosines because we only knew the length of one side. "Maybe we can discover a new law," Trigonometeris said confidently. We looked at the picture of the triangle again. (See Figure 7-3.)

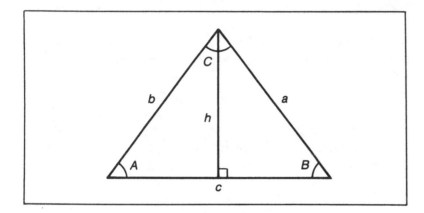

Figure 7-3

"I'm sure you will say we should derive a law called the law of sines, so the sine function won't feel left out," Recordis said. He tried to think of some equations that involved some sines:

$$\frac{h}{b} = \sin A$$

$$\frac{h}{a} = \sin B$$

"We can solve both those equations for h," the professor suggested.

$$h = b \sin A$$

$$h = a \sin B$$

"Now we can say

$$b \sin A = a \sin B$$

We rewrote that equation using the rules of fractions:

$$\frac{b}{\sin B} = \frac{a}{\sin A}$$

"This equation will also be true for all triangles," the king said. We called this the law of sines.

Law of Sines

Let a, b, and c be the lengths of the sides of a triangle, and let A, B, and C be the angles opposite those sides. Then,

$$\frac{a}{\sin A} = \frac{b}{\sin B} = \frac{c}{\sin C}$$

Note that we can include $c/\sin C$ in this equation since the same argument would work in that case. See Exercise 39.

Next, we solved Builder's problem. We let a represent the length of the side opposite the 40° angle and we let b represent the length of the side opposite the 60° angle. Then, according to the law of sines, these two equations must be true:

$$\frac{10}{\sin 80°} = \frac{a}{\sin 40°}$$

$$\frac{10}{\sin 80°} = \frac{b}{\sin 60°}$$

We solved these equations and found $a = 6.527$ and $b = 8.794$.

Builder described another problem: a triangle had two sides equal to 49 and 40, and an angle of 35.256°. Before Recordis could finish reading, Trigonometeris cried, "We will use the law of cosines!"

"I'm afraid that won't work in this case," Recordis said. "Builder says that the 35.256° angle is not between the two known sides. Remember that the law of cosines works when you know the lengths of two sides, and the angle that is between them. Builder says that the 35.256° angle is next to the side of length 49."

"We should be able to use the law of sines, then," Trigonometeris said. He drew a diagram. (See Figure 7-4.) Then he wrote an equation:

$$\sin C = \frac{49 \times \sin 35.256°}{40}$$

$$= 0.7071$$

Trigonometeris happened to recall that $0.7071 = 1/\sqrt{2} = \sin 45°$, so he confidently announced that $C = 45°$. (In Chapter 10 we develop a more

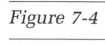

Figure 7-4

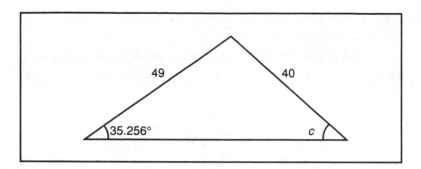

systematic way to determine the size of an angle if the value of its sine is known.)

"That's not right," Recordis argued.

"It has to be right!" Trigonometeris exclaimed. "We have proved that the law of sines works for all triangles."

"The triangle you have drawn (Figure 7-4) is an acute triangle. However, if you had let me finish Builder's letter, I could have told you that Builder's triangle is an obtuse triangle."

Trigonometeris and Recordis were about to have a violent argument when the king realized something. "We know $\sin C = 0.7071$, which is true if $C = 45°$. However, this is also true if $C = 135°$." So, he drew a new triangle. (See Figure 7-5.)

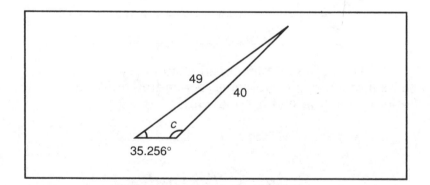

Figure 7-5

"This means that the original specifications we were given were ambiguous," the professor said. "We were given the lengths of two sides of the triangle, and the size of one angle other than the angle between the two given sides. There are two possible triangles that meet those specifications, one obtuse and one acute. If we had not been given the additional information that we were looking for an obtuse triangle then we would have been stuck."

Recordis read the final part of Builder's letter. "The gremlin gave me one final challenge. He asked me to draw a triangle with sides of 40 and 15, with a 35°

angle next to the side of length 40, but not between that side and the side of length 15."

We set up the equation from the law of sines, using C to represent the angle opposite the side of length 40:

$$\sin C = \frac{40 \times \sin 35°}{15}$$

$$= 1.529$$

"No!" Recordis screamed. "That violates the Law of Possible Values for the sine function! We cannot have $\sin C$ greater than 1."

"The gremlin is trying to trick us," the professor guessed. "I bet there is no triangle in existence that meets those conditions."

We drew a diagram and were able to convince ourselves that it is impossible to draw a triangle with a 35° angle next to a side of length 40 which also contained a side of length 15 that was not adjacent to the 35° angle. (See Figure 7-6.)

Figure 7-6

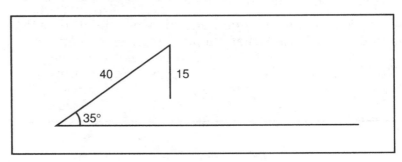

The professor developed a summary of the rules for solving triangle problems, depending on what information you had been given:

Solving Triangles

1.   If you know two angles of a triangle, then you can easily find the third angle (since the sum of the three angles must be 180°).

2.   If you know the three angles of a triangle but do not know the length of any of the sides, then you can determine the shape of the triangle, but you have no idea about its size.

3.   If you know the length of two sides (a and b) and the size of the angle between those two sides (C), then you can solve for the third side (c) by using the law of cosines:

$$c^2 = a^2 + b^2 - 2ab\cos C$$

4. If you know the length of one side (a) and the two angles next to that side (B and C), then you can find the third angle ($A = 180° - B - C$) and then use the law of sines to find the remaining two sides:

$$b = a\sin B/\sin A$$

$$c = a\sin C/\sin A$$

5. If you know the length of the three sides, then use the law of cosines to find the cosine of the angles:

$$\cos C = \frac{a^2 + b^2 - c^2}{2ab}$$

You may find similar expressions for $\cos A$ and $\cos B$. (The professor was beginning to wonder how you could solve for the value of C if you knew the value of $\cos C$. We investigate that problem in Chapter 10.)

6. If you know the length of two sides (b and c) and the size of one angle other than the one between those two sides, then there are four possibilities if the given angle is acute. Suppose you know angle B. Then use the law of sines:

$$\sin C = \frac{c\sin B}{b}$$

a. If $c \sin B/b$ is less than 1 and $b < c$, there are two possible values for C, one obtuse and one acute, and there are two triangles that satisfy the given specifications. This is called the ambiguous case.

b. If $c \sin B/b = 1$, then C is a right angle, and there is only one triangle that satisfies the given specifications.

c. If $c \sin B/b$ is greater than 1, there is no triangle that satisfies the given specifications (since $\sin C$ cannot be greater than 1).

d. If $b > c$, then angle C must be an acute angle. There is only one triangle that satisfies the given specifications.

If the given angle B is obtuse or right, there are two possibilities:

e. If $b > c$, there is one triangle that satisfies the given specifications.

f. If $b \leq c$, there are no triangles satisfying the given specifications. (See Figure 7-7.)

Figure 7-7

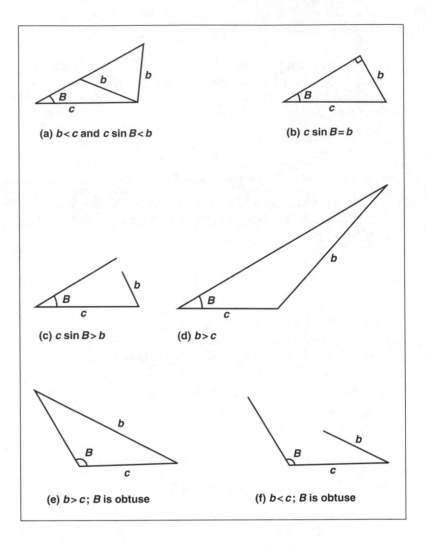

(a) $b < c$ and $c \sin B < b$

(b) $c \sin B = b$

(c) $c \sin B > b$

(d) $b > c$

(e) $b > c$; B is obtuse

(f) $b < c$; B is obtuse

"It can't be too much longer before the bridge is finished," Recordis said. "There can't be very many more problems the gremlin could confront us with."

In Exercises 1 to 10, you are given two sides of a triangle and the angle between those two sides. Calculate the length of the third side.

Exercises

1. 12, 16, 20°
2. 100, 200, 150°
3. 1, 100, 45°
4. 36, 5, 23°
5. 17, 18, 60°
6. 105, 56, 25°
7. 20, 2, 63°
8. 28, 96, 67°

9. 61, 34, 17°

10. 66, 13, 6°

In Exercises 11 to 13, you are given the lengths of the three sides of a triangle. Calculate the three angles.

11. 15, 15, 15

12. 10, 10, 10√2

13. 2, 2, 2√3

14. Consider a triangle that has a 30° angle opposite a side of length 20. One of the sides adjacent to the 30° angle has length 20√3. Calculate the length of the third side.

15–24. Calculate the missing parts of the triangles shown in Figure 7-8.

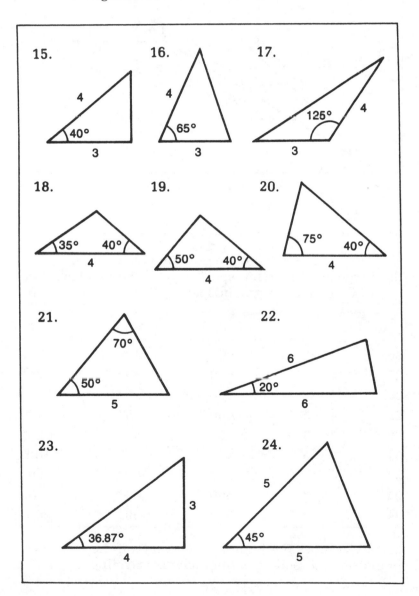

Figure 7-8

25.   Suppose you are piloting an airplane with an airspeed of v in a direction of A north of east. The wind is blowing with a velocity of w in a direction B north of east. Let's put the tail of the wind vector on the tip of the airspeed vector. Then we can draw a new vector that starts at the base of the airspeed vector and ends at the tip of the wind vector. This vector represents the plane's groundspeed—that is, its speed relative to the ground. (See Figure 7-9.) Let s represent the magnitude of the groundspeed vector. Write a formula that expresses s in terms of v, w, A and B.

Figure 7-9

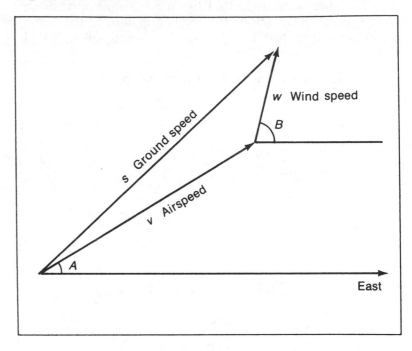

In Exercises 26 to 34, you are given values for the airspeed v, the wind speed w, and the two angles A and B. Calculate the groundspeed (s).

	w	v	A	B
26.	20	600	10°	30°
27.	20	600	10°	120°
28.	20	600	10°	5°
29.	5	400	45°	10°
30.	5	400	45°	40°
31.	5	400	45°	180°
32.	2	500	60°	70°
33.	2	500	60°	0°
34.	2	500	60°	200°

35.   What does the formula say about s if the wind is in the same direction as the plane is traveling $(B = A)$?

36. What does the formula say about s if the wind is blowing in the opposite direction to that the plane is traveling $(B = 180° + A)$?

37. What does the formula say about s if the wind is blowing at right angles to the plane's direction of travel $(B = 90° + A)$?

38. Suppose the plane's groundspeed equals its airspeed, but the windspeed is not zero. Find a formula for $\cos(B - A)$.

39. Show that we may include $c/\sin C$ in the law of sines:

$$\frac{a}{\sin A} = \frac{b}{\sin B} = \frac{c}{\sin C}$$

Suppose that the planets move around the sun in perfectly circular orbits. (See Chapter 5, Exercise 84 for a table that lists the distance from each planet to the sun.) In Exercises 40 to 50, you are given the angle between the planet and the sun as seen from Earth at a particular time. Calculate the distance from Earth to the planet.

40. Mercury 5° 44. Mars 10° 48. Jupiter 160°

41. Mercury 20° 45. Mars 170° 49. Saturn 15°

42. Venus 10° 46. Jupiter 20° 50. Saturn 165°

43. Venus 40° 47. Jupiter 90°

51. Show that this formula is true for any triangle:

$$a = b\cos C + c\cos B$$

This is called a *projection formula*. You can find a similar formula for b and c.

★52. Show that these formulas are true for any triangle:

$$\frac{a+b}{c} = \frac{\cos[\frac{1}{2}(A - B)]}{\sin(\frac{1}{2}C)} \qquad \frac{a-b}{c} = \frac{\sin[\frac{1}{2}(A - B)]}{\cos(\frac{1}{2}C)}$$

These formulas are called *Mollweide's formulas*. You can find similar formulas for $(b + c)/a$, $(b - c)/a$, $(c + a)/b$, and $(c - a)/b$.

★53. Show that for any triangle these formulas are true:

$$\frac{a-b}{a+b} = \frac{\tan[\frac{1}{2}(A - B)]}{\tan[\frac{1}{2}(A + B)]}$$

$$\frac{b-c}{b+c} = \frac{\tan[\frac{1}{2}(B-C)]}{\tan[\frac{1}{2}(B+C)]}$$

$$\frac{c-a}{c+a} = \frac{\tan[\frac{1}{2}(C-A)]}{\tan[\frac{1}{2}(C+A)]}$$

These formulas are called the *law of tangents.*

★54.   Derive Hero's formula. If you have a triangle with sides of length a, b, and c, and we let $s = (a + b + c)/2$, then Hero's formula says that the area of the triangle can be found from this formula:

$$\text{Area} = \sqrt{s(s-a)(s-b)(s-c)}$$

☐55.   Write a program that reads in values for the three sides of a triangle, calculates the area of the triangle, and then calculates the lengths of the three altitudes of the triangle. (An altitude is a line segment perpendicular to one side of a triangle and connecting that side to the opposite vertex.)

☐56.   Write a program that reads in the lengths of two sides of a triangle and the size of the angle between those sides, and then calculates the length of the third side.

☐57.   Write a program that reads in the size of two angles of a triangle and the length of the side between those two angles, and then calculates the lengths of the other two sides.

☐58.   You are lost on an endless, flat plain. Fortunately, you know the exact distances to two radio transmitters at known positions. Find an equation of a line that passes through your position, given this information:

	Transmitter 1	Transmitter 2
x coordinate	$x_1 = 30$	$x_2 = 50$
y coordinate	$y_1 = 40$	$y_2 = 80$
Distance from you	$R_1 = 32$	$R_2 = 18$

To do this, find the distance D between the transmitters:

$$D = \sqrt{(x_2 - x_1)^2 + (y_2 - y_1)^2}$$

Create a triangle with sides of length D, R_1 and R_2. (See Figure 7-10.) After you have found the angle θ_1, note that $d_1 = R_1 \cos \theta_1$. The coordinates of point C are

$$\left[c_x = x_1 + \frac{d_1(x_2 - x_1)}{D}, \quad c_y = y_1 + \frac{d_1(y_2 - y_1)}{D} \right]$$

The equation of the line becomes

$$\frac{y - c_y}{x - c_x} = -\frac{x_2 - x_1}{y_2 - y_1}$$

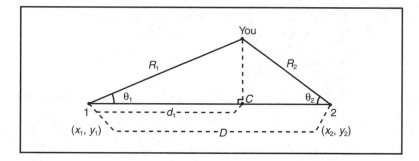

Figure 7-10

8

Graphs of Trigonometric Functions

Builder returned to Capital City the next day riding in his latest invention, a wagon with spring suspension. "The bridge is almost finished," he said cheerfully.

"Then we must plan for the celebration!" Recordis exclaimed.

That evening Builder took us for rides in the wagon. The wagon had a large light on the top so we could see the way. Trigonometeris decided to take a picture of the wagon while the rest of us went for a short ride.

"I have only one problem with the wagon," Builder explained. "As long as the wagon is traveling on nearly level ground, or on ground with only small bumps, everything is fine. However, if the wagon ever hits a large bump. . . ." He was interrupted suddenly when the wagon hit a large bump. The wagon started bouncing smoothly up and down.

"I'm getting seasick," Recordis complained. It was several minutes before the wagon's up-and-down motion began to slow.

"The springs cause the wagon to move up and down like that," Builder explained. "I still need to figure out a way to cause the spring's motion to damp out and come to a stop much more quickly than it does now."

Suddenly we heard an anguished cry from Trigonometeris. "The shutter was stuck open!" he cried. He stood next to the camera tripod and sobbed. "The shutter was open the entire time you were riding in front of the camera," he said sadly.

The professor thought we should develop the picture anyway. We were amazed when we got the picture back from the darkroom. (See Figure 8-1.)

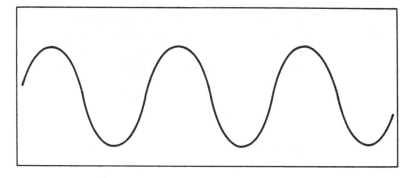

Figure 8-1

"What is that?" Recordis asked in awe.

"I know what happened," the king said. "This is a *time-exposure photograph*. We are seeing the pattern of motion of the light at the top of the wagon. It was too dark for anything but the light itself to show up in the picture."

"We should be able to think of a function that describes that graph!" the professor said excitedly. "When we did algebra we found we were able to understand a curve better if we were able to find a mathematical function that could be represented by the curve."

"We don't know of any function that goes up and down like that!" Recordis complained.

"Let us state the problem more precisely," the professor said. "Our graph always repeats the same pattern. This is the part of the pattern that is always repeated." (See Figure 8-2.)

Figure 8-2

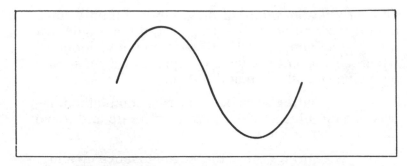

"The graph we are interested in could be formed simply by drawing that one pattern over and over again," the king agreed.

"I have an ingenious idea," the professor said as modestly as she could. "When we have a function that periodically repeats the same pattern, we will call it a *periodic* function, and we will call the length of the pattern the *period* of the function."

The Periodic Function

The professor went on excitedly. "Let's use $f(q)$ to represent our mysterious periodic function. To find the identity of the mysterious function, we will need to find some clues. Let's say p is the period of the function. Then, suppose we know the value of $f(q_1)$ for some particular value q_1. For example, suppose $f(q_1) = \frac{1}{2}$. Then, if we move along the function a distance equal to one period length, we know that the value of the function must be the same:

$$f(q_1 + p) = \frac{1}{2}$$

"Or, in general,

$$f(q + p) = f(q)$$

for any value of q." (See Figure 8-3.)

Figure 8-3

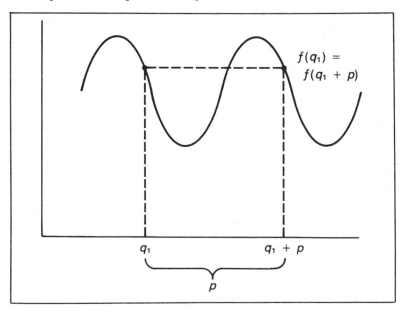

$$f(q_1) = f(q_1 + p)$$

q_1 $q_1 + p$

p

"You would also get the same value for the function if you move along the function a distance of two period lengths," the king said.

$$f(q+2p) = f(q)$$

"Or, the result would be the same if you move a distance of $3p$ or $4p$ or $5p$. . . ." The professor got carried away.

$$
\begin{aligned}
f(q+5p) &= f(q+4p) \\
&= f(q+3p) \\
&= f(q+2p) \\
&= f(q+p) \\
&= f(q)
\end{aligned}
$$

"That is all interesting, but it does not give us a clue to the identity of the mysterious function," Recordis interrupted. "We don't know of any real functions that are periodic."

The professor spent hours trying to think of a periodic function, but she had no success. Recordis doubted that any periodic functions did in fact exist, but he took some consolation from the fact they were working on a problem that did not have anything to do with trigonometry. Finally, he decided to needle Trigonometeris by playing a game.

"I'm thinking of an angle expressed in radian measure," Recordis told Trigonometeris. "The sine of this angle is equal to $1/\sqrt{2}$. Now, you tell me what angle I am thinking of."

"That's easy as *pi*," Trigonometeris said. "The angle is $\pi/4$."

"Wrong!" Recordis exclaimed.

"What?" Trigonometeris screamed. "That has to be right!" But Recordis resolutely shook his head.

"I admit that sin $(\pi/4) = 1/\sqrt{2}$," he said. "But that's not the angle I am thinking about."

"I see," Trigonometeris suddenly realized. "This is a trick question. The angle is $3\pi/4$."

"Wrong again!" Recordis said. "I bet you'll never guess it!"

Trigonometeris started screaming that Recordis still did not understand trigonometric functions, but then he realized another possibility. "The angle could be $(2\pi + \pi/4)$." Recordis shook his head. "It could be $(4\pi + \pi/4)$," Trigonometeris guessed. Again Recordis shook his head.

"This is impossible for me to guess!"

Trigonometeris cried. "We know that an angle remains exactly the same if you add 2π to it. Therefore,

$$\sin\frac{\pi}{4} = \sin\left(\frac{\pi}{4} + 2\pi\right)$$
$$= \sin\left(\frac{\pi}{4} + 4\pi\right)$$
$$= \sin\left(\frac{\pi}{4} + 6\pi\right)$$
$$= \cdots$$

"Or for any value of x, we know that

$$\sin x = \sin(x + 2\pi)$$
$$= \sin(x + 4\pi)$$
$$= \sin(x + 6\pi)$$
$$= \cdots$$

"I bet you never would have guessed my angle," Recordis said. "I was thinking of ($\pi/4$ + 2,316,978π)."

"Will you two be quiet!" the professor cried. "I am trying to think of a periodic function." She suddenly noticed the string of equations that Trigonometeris had written on the board.

"That's it!" she realized. "The sine function is a periodic function! Whenever you increase x by 2π, then the value of the sine function remains the same. Therefore, $\sin x = \sin(x + 2\pi)$ and therefore the sine function is a periodic function with a period length of 2π."

"Of course!" Trigonometeris realized. "Why didn't I think of that!"

"That still doesn't mean that the sine function is the correct function to describe the motion of the spring-driven wagon," Recordis cautioned. (He was miffed that this had turned into a trigonometry problem after all.)

"There is only one way to proceed," Trigonometeris said. "We must make a graph of the function $y = \sin x$ to see what it looks like."

"It takes a lot of work to draw a graph of a function!" Recordis complained, "and we know who ends up doing most of the work around here. To draw this graph I will need to look carefully at the table of values and draw a lot of dots. Then I will need to see if I can connect the dots with a smooth curve."

"I will be extraspecial nice to you if you do this for me," Trigonometeris promised. "I can already tell you one point on the graph: ($x = 0$, $y = 0$) will be a point, since $\sin 0 = 0$."

"Wait a minute," Recordis said. "First we must figure out the vertical scale and the horizontal scale of the diagram."

"The vertical scale will be easy." Trigonometeris said. "We know that the sine function never reaches a value greater than 1, and it never reaches a value smaller than –1. The horizontal scale will be harder, since we will want the graph to cover all possible values of x. Therefore, we must start at x equals minus infinity and continue until x equals plus infinity."

Recordis fainted.

"It will be much easier than that!" the king said. "We only need to draw the graph for $x = 0$ to $x = 2\pi$. Because the function is periodic, we know it will always repeat the same pattern."

The Graph of the Sine Function

Recordis revived and set to work. Trigonometeris read off the first few entries from the sine table.

Degrees	Radians	Sin
1	0.01745	0.01745
2	0.03491	0.03490
3	0.05236	0.05234
4	0.06981	0.06976
5	0.08727	0.08716

Very carefully, Recordis put a dot on the diagram that matched each of these points. (See Figure 8-4.) The

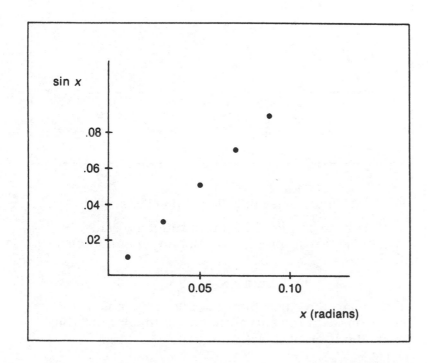

Figure 8-4

whole process took a long time. However, as he added more and more dots, it became clear that the graph of the sine function was a smooth, graceful curve.

"It's beautiful!" Trigonometeris said in awe as the picture grew.

Recordis labored over the diagram for hours. We could see the curve was approaching a dramatic plateau as x approached $\pi/2$ and y approached 1. (See Figure 8-5.) Recordis collapsed with exhaustion after reaching this far.

Figure 8-5

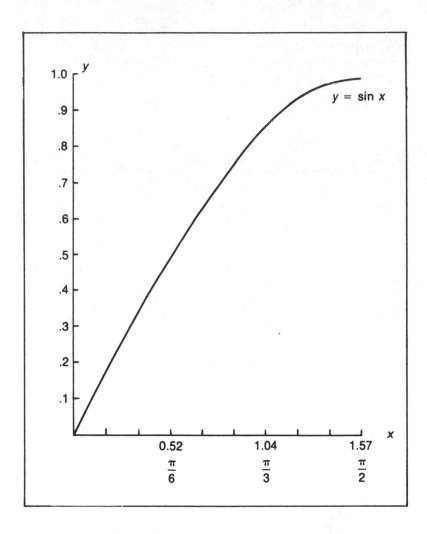

"We must press on!" Trigonometeris cried.

"I have an idea," Recordis suddenly said cheerfully. "I will not have to plot any more points at all. Since

$$\sin(\pi - x) = \sin x$$

"that means the curve from $x = \pi/2$ to $x = \pi$ is just the mirror image of the curve from $x = 0$ to $x = \pi/2$." (See Figure 8-6.)

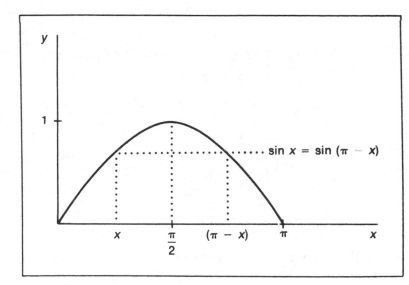

Figure 8-6

Recordis quickly completed the curve from $x = \pi/2$ to $x = \pi$. Then he announced his next idea. "The curve from $x = \pi$ to $x = 2\pi$ will be the same as the curve from $x = 0$ to $x = \pi$, except it will be turned upside down. We know that the angle $2\pi - x$ is the same as the angle $-x$, and since $\sin(-x) = -\sin x$, it follows that $\sin(2\pi - x) = -\sin x$." (See Figure 8-7.)

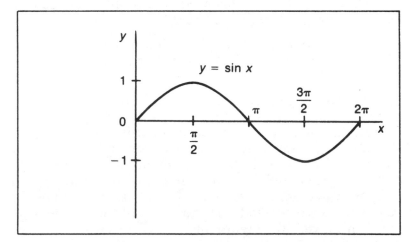

Figure 8-7

"We have completed the entire pattern!" the professor said enthusiastically. "Now it will be easy to draw the entire curve, since we merely need to repeat that same pattern many times." (See Figure 8-8.) Note that the appearance of the curve does change when the scale of the diagram is changed.)

We all stared in admiration at the completed graph of the curve $y = \sin x$. "It's a very elegant curve," Recordis agreed. "I had never realized that trigonometry could be that artistic."

"And it is exactly the curve we need to describe the motion of the bouncing wagon!" Trigonometeris said. Upon close inspection, we could see that the

Figure 8-8

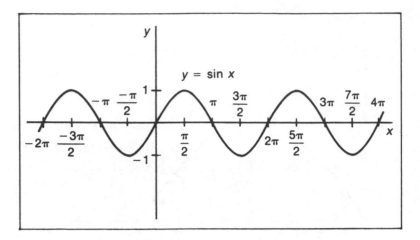

shape of the curve in the time-exposure photograph of the wagon was exactly the same as the shape of the curve $y = \sin x$.

"This result does indeed suggest that trigonometry is much more versatile than we had imagined," the professor said. "It seems that the function $y = \sin x$ can describe the motion of objects that oscillate back and forth, such as objects driven by springs. Originally we developed trigonometry to help us solve problems relating to triangles, but this particular problem doesn't have anything to do with triangles."

"We should make graphs of the other trigonometric functions," Trigonometeris said. "Let's make a graph of the curve $y = \cos x$."

The Graph of the Cosine Function

Recordis panicked at the thought of having to draw a whole curve again, but then he suddenly realized an identity that would help. "Since $\cos x = \sin (\pi/2 - x)$, it seems to me that the cosine function graph will have exactly the same shape as the graph of the sine function—the only difference is that it will be shifted a bit." We knew that $\cos 0 = 1$, and cos $(\pi/2) = 0$, so we guessed that the graph of the cosine function looked like Figure 8-9.

Figure 8-9

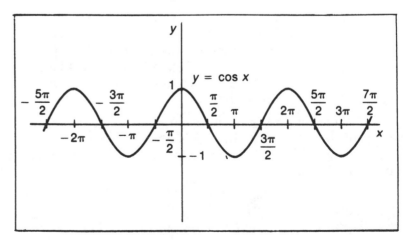

The graphs of the other trigonometric functions did not seem to be as interesting. We made a graph of the function $y = \tan x$. That function was difficult to draw because the value of $\tan x$ approached plus infinity as x approached $\pi/2$, and it approached minus infinity as x approached $-\pi/2$. (See Figure 8-10.)

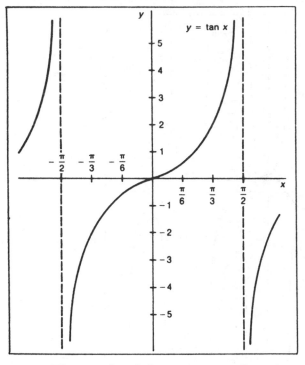

The graph of the cotangent function was simply a shifted version of the tangent function graph. (See Figure 8-11.)

Figure 8-11

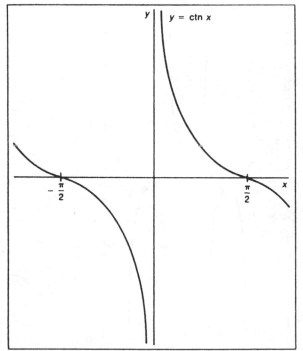

Graphs of the Secant and Cosecant Functions

The graphs of the secant function and the cosecant function were even stranger. (See Figures 8-12 and 8-13. Note that each function is drawn on the same graph as its corresponding reciprocal function.)

Figure 8-12

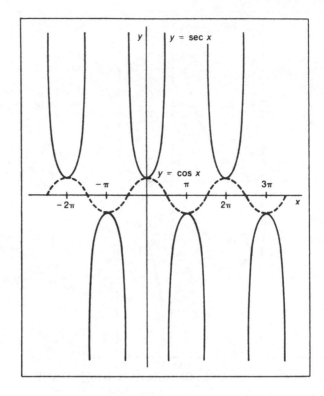

Figure 8-13

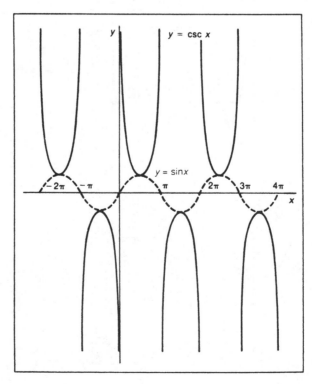

We were all exhausted by the close of the evening, but we were all excited by the accomplishments we had made that day. The next day Builder showed us another invention he had developed while he had been working on the bridge. "I call it *alternating current electricity*, or AC for short," he said proudly. He showed us a device consisting of a steam-driven rotor inside a large magnet connected to a pair of wires. "This is an *electrical generator*," he explained. "When the rotor turns, the generator creates electricity that flows through the wires." He turned the generator on.

Recordis looked closely at the wires. "I don't see any electricity flowing," he said.

"You can't see the electricity itself," Builder said. "But it does help to be able to see the pattern of the current. In alternating current, the electricity sometimes flows one way, then it turns around and flows in the opposite direction. Then it turns around again. I have designed the generator so that each complete turnaround takes $\frac{1}{60}$ of 1 second. To see the pattern of the current, I invented a machine that I call an *oscilloscope*." He showed us a device that looked like a television screen connected to a bunch of knobs. He connected the oscilloscope to the wires coming from the generator, and then he turned it on. We received one of the biggest shocks of our entire lives. (See Figure 8-14.)

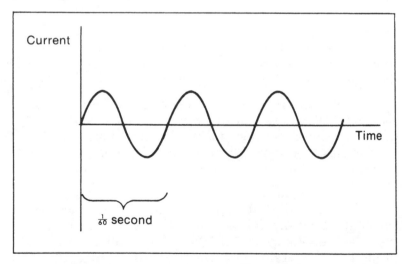

Figure 8-14

"That's a sine curve!" we all gasped in astonishment.

"You mean you recognize that curve?" Builder asked us in surprise. "That curve describes alternating current electricity. I can change the height of the curve by increasing the amplitude of the current." Builder turned a dial and the shape of the sine curve changed. (See Figure 8-15.)

Figure 8-15

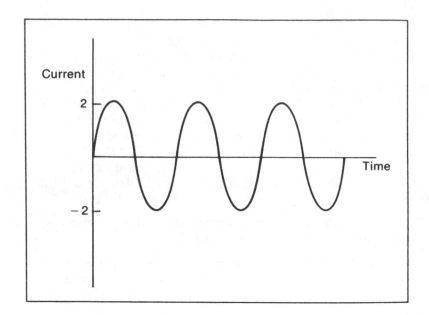

Figure 8-16

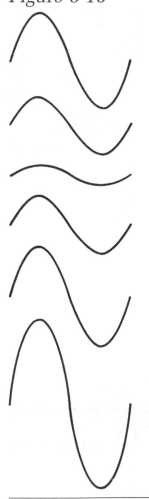

"I guess this curve represents the function $y = 2 \sin t$," the professor suggested. "You have taken the entire sine curve and multiplied every value by 2." (We used t to represent time because the electric current was a function of time.)

Builder showed us that he could adjust the dial to create sine curves of different heights. We decided that the general form for the function describing the current was

$$y = A \sin t$$

where A represented the *amplitude* of the sine function. We drew several different sine curves with different amplitudes. (See Figure 8-16.)

"We have a problem," Recordis suddenly realized. "We know that the sine function has a period of 2π. However, Builder told us that the alternating current has a period of $\frac{1}{60}$ seconds. Therefore, the function $y = A \sin t$ cannot represent the current."

We puzzled over this problem. "We need a way to adjust the period of a sine function," the professor said. She drew several sine functions with different periods. (See Figure 8-17.) "We put the letter A in front of the function $A \sin t$ to allow us to adjust the amplitude," the professor said. "So there must be some place in the function where we could put another letter that would allow us to adjust the period."

The king had an idea. "The function $\sin t$ has a period of 2π, since the sine function runs through a complete pattern every time t runs from $t = 0$ to $t = 2\pi$. Therefore, if we created the function $y = \sin(2\pi t)$, we should see a period of 1."

"I have an idea," Trigonometeris said. "Suppose we want to have a sine function with a period of 3. Then we could use the function

$$y = \sin\left(\frac{2\pi t}{3}\right)$$

We could verify that when t ran from 0 to 3, then $2\pi t/3$ ran from 0 to 2π, and therefore the function $\sin(2\pi t/3)$ ran through a complete cycle.

"I see how to do it in general," the professor said, pleased that she was the one who was always able to see how to state a problem in general terms. "Let's use the letter p to represent the length of the period. Then the function

$$y = \sin\left(\frac{2\pi t}{p}\right)$$

will have a period of length p."

"We can say that the function $y = A\sin(2\pi t/p)$ has a period of length p and an amplitude of A," Trigonometeris added.

"I also find it useful to measure the *frequency* of the current," Builder said. "The frequency tells you the number of cycles that occur each second. If the period

Frequency

Figure 8-17

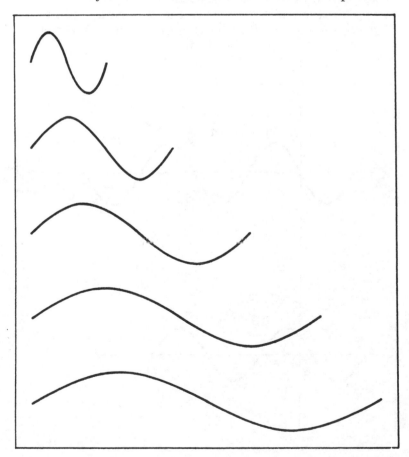

is p and the frequency is f, then $f = 1/p$. For example, when the alternating current has a period of $\frac{1}{60}$ second, then it has a frequency of 60 cycles per second." (One cycle per second is called 1 hertz (Hz), so a frequency of 60 cycles per second equals 60 hertz.)

"We can easily write a sine function that has a period of f," the professor said.

$$y = A \sin(2\pi f t)$$

"Do we have to keep writing that 2π all the time?" Recordis asked.

"I have an idea," the professor said. "Let's define something that I'll call the *angular frequency*, represented by ω. We will define angular frequency as

$$\omega = 2\pi f$$

"Then, a sine function with angular frequency ω can be written as

$$y = A \sin(\omega t)$$

"Why do you use the letter w to represent angular frequency?" Recordis asked.

"That's not a w!" the professor responded. It's a Greek letter called *omega*. It just looks a bit like a w."

"I also have a knob that can shift the entire curve," Builder said. "I refer to that as changing the *phase* of the curve." Builder demonstrated two different curves with the same amplitude and frequency but different phases. (See Figure 8-18.)

Phase

Figure 8-18

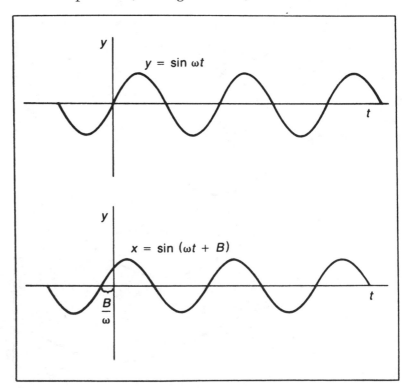

"We can easily describe sine functions with different phases," Trigonometeris said. "All we need to do is include a letter that allows us to adjust the starting point of the function." He suggested that we write the function like

$$A = \sin(\omega t + B)$$

"The term B/ω will measure the phase of the curve. In the examples we have done so far, the phase B/ω is zero."

We wrote down the general formula for a sine curve.

General Formula for a Sine Curve

$$y = A \sin(\omega t + B)$$

where A = amplitude

ω = angular frequency

(in radians per second)

$f = \omega/2\pi$, frequency (in cycles per second)

$p = 1/f = 2\pi/\omega$, period (in seconds)

B/ω = phase

Builder adapted the screen of the oscilloscope into a graphing calculator, which was versatile enough that we could enter the formula for one or more functions and see the graph. (If you have a graphing calculator available, it will be a big help in visualizing the graphs of trigonometric functions. Another way to create graphs is with a computer spreadsheet. The appendix discusses some examples of how to do this.)

Solving an Equation with a Graphing Calculator

"We can use graphs to help solve certain kinds of equations," the professor suggested. "Suppose we need to find the values of x that solve this equation:"

$$\sin x = \sin 3x$$

"If we draw a graph of $\sin x$ and $\sin 3x$, the solutions to the equation will be at the points where the curves cross."

With our graphing calculator we graphed both $y = \sin x$ and $y = \sin 3x$. (See Figure 8-19.) "I can see seven points where the two curves cross between $x = 0$

Figure 8-19

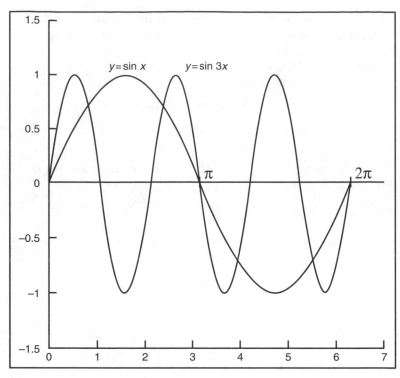

and $x = 2\pi$," Trigonometeris said. "They cross at $x = 0$, $x = \pi$, and $x = 2\pi$. We can verify that these points are solutions since $\sin 0 = \sin(3 \times 0) = \sin(\pi) = \sin(3\pi) = \sin(2\pi) = \sin(6\pi) = 0$. However, the other four solution points are not as obvious."

"I wish we could zoom in on one of the crossing points and determine its exact coordinates," Recordis said wistfully. Builder told us his graphing calculator could zoom in on a point. Figure 8-20 zooms in on one

Figure 8-20

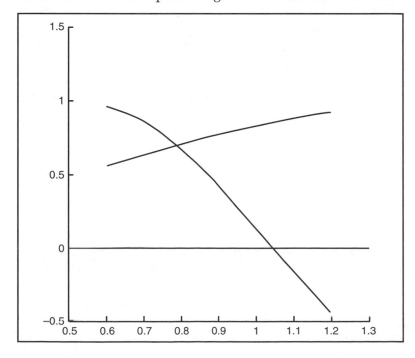

of the solutions. We can see that the value of x at the solution is somewhere between 0.7 and 0.8. We zoomed in some more (Figure 8-21) to find that the solution is between 0.785 and 0.786. Further zooming led us to the

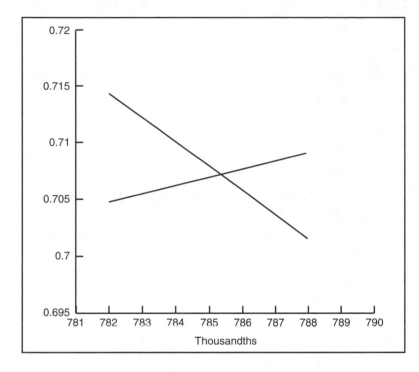

Figure 8-21

value 0.7854, and we began to expect that the solution was in fact $\pi/4$. "It will be hard to find an exact solution from the graphing method," Recordis said, "but we should be able to find a very close approximation."

That evening Trigonometeris dreamed about the new uses he would find for sine curves. "At last we have found the world's most beautiful shape," he thought to himself.

Identify the amplitude, frequency, and angular frequency of the curves in Exercises 1 to 6.

Exercises

1. $9.8\sin(2x + 2)$

2. $\sin(10x + 5)$

3. $5\cos(\pi x)$

4. $\pi\cos(x/\pi)$

5. $100\sin(x/100)$

6. $4.5\cos(50x + 16)$

For Exercises 7 to 10, write the equation of a sine function that has the period indicated.

7. $\frac{1}{2}$

8. 16

9. $\pi/2$

10. $\pi/4$

Draw graphs of the functions in Exercises 11 to 17.

11. $y = \sin x + \sin(x + \pi)$

12. $y = (\sin x + |\sin x|)/2$

13. $y = \sin^2 x$

14. $y = x + \sin x$

15. $y^2 = \sin^2 x$

16. $y = (\sin x)(\sin x/10)$

17. $\sin^2 x + \cos^2 x$

★18. Draw a careful sketch of one arch of the curve $y = \sin x$. Estimate the value of the area of the arch.

□19. Write a program that draws a graph of a sine curve. Design the program so that you may change the amplitude, frequency, and phase of the curve, and make it possible to change the scale of the diagram so you may have either a close up view of one arch of the curve or else a wide angle view of several arches. Once you have written the program, you may modify it to draw graphs of related curves, such as those given in Exercises 11 to 17 or Figures 9-2 to 9-4.

Builder sent us word that the long-awaited day had at last arrived: The Raging River bridge was completed! We hurriedly packed the balloon and sailed to the bridge site. The triangular bridge supports glistened in the sunlight.

"The gremlin will have no choice but to admit defeat now!" Recordis said joyfully.

The big celebration was planned for the next day. That afternoon we relaxed by renting a rowboat on nearby Ripply Lake. For most of the afternoon the water was very calm. However, our tranquility was shattered when a noisy speedboat sped by. It left a chain of rolling waves in its wake that struck our boat.

"I have a feeling I've gone up and down like this before," Recordis said.

"Hold on tight!" the professor cried as the boat bounced on top of the waves. However, Trigonometeris did not heed her. He was leaning over the side of the boat staring at the waves. Just as the professor feared, the boat suddenly lurched and he was thrown overboard.

"Save him!" the king cried.

Recordis threw a life preserver to Trigonometeris, and we managed to pull him back on the boat. Trigonometeris was too excited to notice that he was sopping wet. "Did you see the shape of those waves?" he exclaimed. "They look like sine curves!"

We instantly realized the significance of what he had said. "Let's investigate the properties of waves," the professor said quickly. "Let's use y to represent the height of the water at a particular time t. Then I bet

$$y = \sin t$$

"We felt the boat go up and down, just like the sine function."

"We're forgetting something," the king said. "It would help to be able to describe the nature of the wave at every single location on the lake. The equation $y = \sin t$ only describes the wave at one location."

"Suppose we look at the entire lake at one particular time," Trigonometeris said. "Then it is clear that

$$y = \sin x$$

where y gives the height of the wave at a distance x away from the shore."

"But that equation does not take into account changes in the wave with time!" the professor protested.

"Is there any way we could write one function that could describe both the wave movement with time and its variation at different points on the lake?" the king asked in puzzlement.

"We could try to make a sine function that includes both x to measure position and t to measure time," Recordis suggested.

$$y = \sin(x - t)$$

"If we look at that function from a fixed location ($x = x_0$), then we will see that the water at that location goes up and down with time. Or, if we look at that function at a fixed time ($t = t_0$) then we will find that the pattern of water at different locations looks like a sine curve."

"Ingenious!" the professor realized. "Represent a wave by a sine curve that is a function of both space x and time t."

At that moment the boat was rocked by a new set of waves. However, these waves were much less violent than the first set of waves.

"To make our wave function more general, we must put a letter in front of the sine to represent the amplitude," Trigonometeris said. "It is clear that not all waves are created equally, so we should be able to represent waves of different amplitudes."

We wrote down a new wave function, using A to represent the amplitude of the wave:

$$y = A\sin(x - t)$$

The king watched the waves move across the water. "We must find a way to represent the velocity (or speed) of the waves," he realized. We let v represent the velocity of the wave. Then, after some experimentation, we found that our new wave function should look like

$$y = A\sin(x - vt)$$

We will use the term crest for the point at the top of a wave.

"A crest of this wave will move with velocity v," the professor realized with satisfaction. "If the x coordinate of the crest is x_{crest}, then y must have its maximum value—that is, $y = A$:

$$A = A\sin(x_{crest} - vt)$$
$$1 = \sin(x_{crest} - vt)$$
$$x_{crest} - vt = \frac{\pi}{2}$$
$$x_{crest} = \frac{\pi}{2} + vt$$

From this equation we can clearly see that with every increase in t by 1, the position of a crest (x_{crest}) will increase by v."

"There is one more thing we need to take into account," Trigonometeris said. "With some waves, there is a very long distance between each crest. Other times the crests are very close together. We need to include a way to adjust for that distance."

We put a k in the middle of the sine function:

$$y = A\sin[k(x - vt)]$$

"I can see in principle how k allows us to adjust the distance between crests," Recordis said, "but I would feel more comfortable if we could translate it into something more meaningful."

"We should measure the distance between the crests and call that the *wavelength*," the professor said. "I like the Greek letter *lambda* λ, so let's use λ to represent the wavelength." (See Figure 9-1.)

Figure 9-1

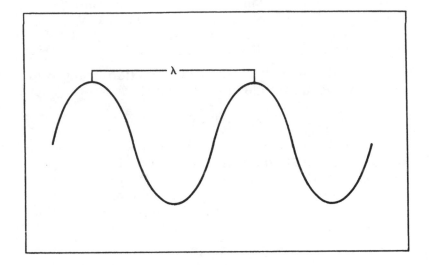

"Can we calculate an expression for the wavelength in terms of *k*?" the king wondered.

"Let's suppose that one crest occurs at the point x_1, and the next crest occurs at the point x_2," Trigonometeris said. "Then, $\lambda = x_2 - x_1$. During the course of one wave, the argument of the sine function, which is $k(x - vt)$ in this case, must increase by 2π. Therefore,

$$k(x_1 - vt) + 2\pi = k(x_2 - vt)$$

We rewrote that as

$$kx_1 - kvt + 2\pi = kx_2 - kvt$$

The two $-kvt$'s canceled out:

$$kx_1 + 2\pi = kx_2$$
$$2\pi = kx_2 - kx_1$$
$$= k(x_2 - x_1)$$
$$= k\lambda$$
$$\lambda = \frac{2\pi}{k}$$

"This formula tells us how to calculate λ if we know *k*!" the professor said, "and I just thought of an interpretation for *k*. *k* will tell us the number of waves in a distance of 2π. We will call it the *wave number*."

"When I'm being tossed around on the boat, I'm not that interested in the wavelength," Recordis said. "I am really much more interested in the number of times I am bounced up and down each second."

"We can calculate the number of crests that pass a given point in one unit of time," the professor said. "We will call that quantity the *frequency* of the wave."

"I know how to calculate frequency," Trigonometeris said. "Each wave crest is a distance of λ apart, and each crest is moving with a speed *v*. If *v* = 1 meter per second, and λ = 1 meter, then we will be hit by one wave each second. If λ = 2 meters, then we will only be hit by one-half wave each second. In general, the frequency will be

$$f = \frac{v}{\lambda}$$

where *f* is the number of waves that hit us each second."

We also found it convenient to define a quantity called the angular frequency (again represented by ω): ω = 2π*f*. Then we summarized our results for waves. (We called this type of wave a *harmonic wave*.)

★*Harmonic Waves*

A one-dimensional harmonic wave can be represented by a function like

$$y = A \sin (kx - \omega t + B)$$

where x = location

t = time

A = amplitude

k = wave number, the number of waves in a distance of 2π units

λ = 2π/k, wavelength (distance between crests)

ω = angular frequency

f = ω/2π, frequency (number of waves that pass a point per unit time)

p = 1/f = 2π/ω, period (amount of time it takes a wave to pass a point)

v = ω/k = λf, velocity of the wave

B = a parameter that allows you to adjust the initial phase of the wave

(Water waves are not the only examples of harmonic waves. Sound and light are made up of waves. The function just given works exactly for waves that are one-dimensional, such as waves on a string. Water waves occur on a two-dimensional surface, and sound and light waves occur in three-dimensional space. The equations for those types of waves are more complicated, but the principles of wavelength and frequency are the same.)

We rowed the boat back to shore and returned to our camp. Some of the musicians were tuning their instruments in preparation for the gala celebration the following day.

"What's that sound?" the professor asked. A piercing hum was coming from the pavilion. We went inside and found a worker checking some tones by tapping a fork-shaped metal device.

"I call this a *tuning fork*," the tuner explained. "When I tap the fork, it makes a sound with a very precise pitch."

The professor watched the tuning fork closely. "It vibrates back and forth after you hit it!" she exclaimed. "But I wonder why vibrations make sound?"

"Whatever sound is, it must be able to travel through air," the king said thoughtfully.

"We found that sine functions describe back-and-forth motion," Trigonometeris said helpfully.

"Will you stop trying to get trigonometry involved in everything!" Recordis protested.

★ Sound Waves

"I bet I know what happens when the tuning fork vibrates," the professor said. "The fork must push the little air molecules back and forth. Those molecules must push against some of the other air molecules. I bet a chain reaction is started, until finally the little molecules near our ears are pushed, and then our eardrums start to vibrate." Her eyes widened. "I bet sound travels as a wave! We saw how waves travel in water. I bet that sound consists of invisible waves that travel through air!"

We were skeptical, but we conducted several experiments to see if sound behaved like waves. We struck several tuning forks. We found that the forks that vibrated faster emitted sounds of higher pitch. We guessed this meant that a high-pitched sound wave has a higher frequency than a low-pitched sound wave. (The wave itself consists of regions of higher density air alternating with regions of lower density air.) We conducted a careful experiment to measure the speed of sound and found that v equals about 339 meters per second (about 760 miles per hour). The professor's wave theory received even more support when we discovered that a tuning fork that emitted a standard A tone had a frequency of 440 cycles per second (or 440 hertz).

"Now we can calculate the wavelength of a sound wave corresponding to the A tone," the professor said eagerly. "We know that $v = 339$ meters per second, we know $f = 440$ cycles per second, and we know that for any wave $v = \lambda f$. Therefore,

$$\lambda = \frac{v}{f}$$

$$= \frac{339 \text{ meters/second}}{440 \text{ cycles/second}}$$

$$= 0.77 \text{ meters}$$

The professor set up a row of tuning forks and found that she could make sounds of many different pitches by tapping forks of different sizes.

"I hate to disillusion you, but none of these sounds are very much like music," Recordis said. "They have different pitches, just like musical notes have different pitches. However, there is something different about the quality of sound that comes from a musical instrument—it just doesn't sound like the sound that comes from a tuning fork."

The professor was crestfallen, but she realized Recordis was right. "It will be more complicated to analyze sound than I had thought," she said.

That evening we built a campfire. The professor was still trying to figure out how to analyze sound. Trigonometeris decided to amuse himself by drawing different types of sine curves with different frequencies. A playful idea occurred to him. "I wonder what happens if you try to add together two sine functions with different frequencies," he mused. He set up a new function:

$$y = \sin t + \sin(2t)$$

"The first function has a frequency of $1/2\pi$. The second function has a frequency of $1/\pi$." He carefully sketched the graph of this function. (See Figure 9-2.)

"That is an interesting graph," the professor noted. "You can still see the effects of each individual sine curve in the combined curve. I can't see that it is good for anything, though."

Trigonometeris tried some more drawings where he added together different sine curves. (See Figures 9-3 and 9-4.)

★*Adding Sine Functions of Different Frequencies*

Figure 9-2

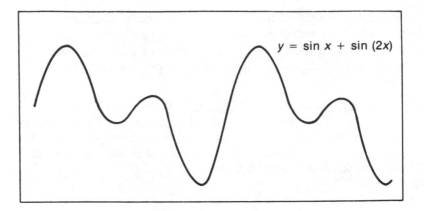

$$y = \sin x + \sin (2x)$$

Figure 9-3

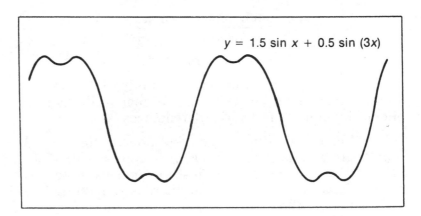

$$y = 1.5 \sin x + 0.5 \sin (3x)$$

Figure 9-4

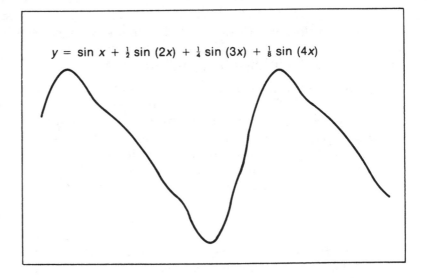

$$y = \sin x + \tfrac{1}{2} \sin (2x) + \tfrac{1}{4} \sin (3x) + \tfrac{1}{8} \sin (4x)$$

★*Standing Waves in Guitar Strings*

The king took out his guitar to play some campfire songs. The rest of us were gently lulled to sleep by the relaxing melodies. However, the professor was staring at the king's fingers. Whenever he plucked a string, a musical tone sounded. The professor noticed that each string vibrated after being plucked.

"I wonder if we can mathematically describe the vibration of a string," she thought. She drew a sketch of a string of length L, fastened down at both ends. "I bet the string would look like this if we could stop it for one instant." (See Figure 9-5.) "We should be able to represent this pattern as a sine curve, like

$$y = A \sin kx$$

"Now all we need to do is track down the value of k," she continued thinking. "We know that y must

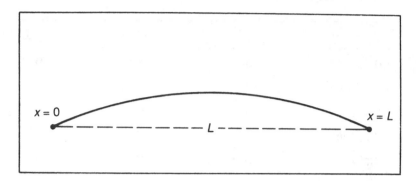

Figure 9-5

equal 0 when $x = 0$. Fortunately, that will be true, no matter what the value of k. We also know that y must be 0 when $x = L$, because the string is tied down at the other end. That means

$$0 = A \sin kL$$

$$0 = \sin kL$$

"I bet that means that $kL = \pi$. Therefore, $k = \pi/L$. Now we need to represent the fact that the string moves with time." She decided to write the function for the string with this formula:

$$y = A \sin(kx) \sin(\omega t)$$

where y represents the distance away from its resting position at time t for a point on the string at a distance x from the end (Figure 9-6). This function seemed to the professor to have the desirable properties: at $t = 0$, $t = \pi/\omega$, $t = 2\pi/\omega$, and so on, the value of y was 0 for all points on the string, meaning that the string was momentarily back in its resting position. At $t = \pi/2\omega$,

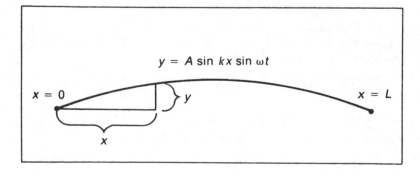

y = A sin kx sin ωt

Figure 9-6

$t = 2\pi/\omega + \pi/2\omega$, $t = 4\pi/\omega + \pi/2\omega$, and so on, the string was at its maximum positive displacement; at $t = 3\pi/2\omega$, $t = 2\pi/\omega + 3\pi/2\omega$, and so on, the string was at its maximum negative displacement. At $x = 0$ and $x = L$, the value of y was 0 at all times, because the two ends were tied down.

She created a table showing values of $y = A\sin(kx)$ $\sin(\omega t)$, where $k = \pi/L$:

	$x = 0$	$x = L/2$	$x = L$
$t = 0$	0	0	0
$t = \pi/(2\omega)$	0	A	0
$t = \pi/\omega$	0	0	0
$t = 3\pi/(2\omega)$	0	$-A$	0
$t = 2\pi/\omega$	0	0	0

She excitedly woke everyone else to announce her discovery.

"That looks like a wave!" Trigonometeris said when he saw her function. "Only this is not the same as the moving waves we discussed earlier. This type of wave remains standing in one place." We decided to call this type of wave a *standing wave*.

"We can calculate the wavelength," the professor said. "Since $k = \pi/L$ and $\lambda = 2\pi/k$, therefore, $\lambda = 2L$."

"If we knew the velocity of the wave in the string, we could find the frequency of the vibration, using the formula

$$f = \frac{v}{\lambda}$$

the king said. The string was $L = 0.5$ meters long, so the wavelength was $\lambda = 2L = 1$ meter. We found that the wave velocity in the string was 440 meters per second. Therefore, we found that $f = 440$ cycles per second = 440 hertz. We found that the string did indeed generate a sound of frequency 440 hertz (which is the musical note A above middle C) when it was plucked.

The professor was glad for the chance to show off, so she recounted every detail of her development of the standing-wave formula.

"How did you calculate the value of k?" Recordis asked.

"It was easy," the professor said. "I know that y must be 0 when $x = L$, so from the formula

$$0 = \sin kx$$

"I could clearly see that $kL = \pi$."

"But we know that sin kx will also be zero when $kx = 2\pi$, $kx = 3\pi$, $kx = 4\pi$, and so on," Recordis pointed out.

The professor suddenly stopped.

"That means there are many possible values for k," the king said. "We know that any value of k such that $k = n\pi/L$, where n is an integer, will be allowable."

"That means there are many possible values for the wavelength," Trigonometeris said. (See Figure 9-7.)

$$\lambda = \frac{2L}{n}$$

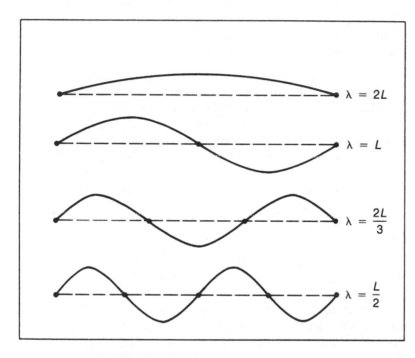

Figure 9-7

"And, there are many possible values for the frequency," Recordis said.

$$f = \frac{vn}{2L}$$

The professor was quite embarrassed. She had thought she had waves figured out perfectly, but now it turned out she could not even determine the frequency of the vibration for certain. "But we already found that the guitar string made a sound of frequency 440 hertz," she protested.

"I bet it also makes sounds at some of these other frequencies," Trigonometeris said. Builder quickly constructed a frequency detector, and we found that the string also emitted sounds with frequencies of 880 hertz, 1320 hertz, 1760 hertz, and some higher frequencies. All the frequencies were multiples of 440 hertz.

"This is getting very complicated," Recordis said. "One guitar string generates sounds at so many different frequencies. We can see from Builder's frequency detector that each frequency has a different amplitude."

"This is a bit like the drawings I was doing this afternoon," Trigonometeris said. "I drew functions that consisted of several sine functions of different frequencies added together."

Suddenly the professor's jaw dropped open. "I know what makes music!" she cried. "A tuning fork generates a sound of one pitch, but it doesn't sound like music. A guitar string generates sound of many frequencies that are all multiples of one fundamental frequency—and the guitar sounds like music!"

★Music

We conducted further investigations that demonstrated the professor was right. A musical note consists of a base frequency, the *fundamental frequency*. The note also consists of a mixture of sounds of higher frequencies that are multiples of the fundamental frequency. These higher frequencies are *harmonics*. The quality of a musical tone is determined by the exact mixture of the harmonics involved. Two notes of the same pitch coming from different musical instruments will sound different because of a different pattern of harmonics.

The celebration the next day was very festive. The king cut the ribbon on the bridge, and Builder drove his wagon across for the first time. The band played triumphant marches. That evening we returned to Capital City, where the Royal Symphony played a special concert in honor of the new bridge. All during the concert the professor was carefully observing every corner of the concert hall thinking of ways to improve the acoustics now that she understood about sound waves. However, just as the concert was approaching its dramatic conclusion, we were interrupted by a loud thunder clap. Before us stood none other than the gremlin!

"I'm surprised you dare show your face, you vile creature!" the king said defiantly. "We built the bridge over Raging River in spite of your threats!"

The Threat of the Terrible Flood

The gremlin just laughed. "So you did. However, you will now face a more difficult challenge. I have sent a large wave that will engulf the town of Peaceful Bay."

"That is no problem!" Builder said. "I can design a breakwater that will stop the flood!"

"Indeed you can," the gremlin laughed as he vanished. "But how will your design reach Peaceful Bay in time?"

- Any periodic function can be expressed as the sum of sine curves of different frequencies. This result is known as the *Fourier theorem*.

Identify the wavelength, frequency, and velocity of the waves given by the wave functions indicated in Exercises 1 to 6.

1. $y = A\sin(x - t)$

2. $y = A\sin(2x - 3t)$

3. $y = A\sin(2\pi x - 2\pi t)$

4. $y = A\sin(1.14x - 3.48t)$

5. $y = A\sin(2x - 2t)$

6. $y = A\sin(8000x - 2400t)$

The velocity of light (and all electromagnetic waves) is 3×10^8 meters per second. Exercises 7 to 14 give frequencies in hertz of some electromagnetic waves. Calculate the wavelength of each wave.

7. 1.4×10^{20} (X ray)

8. 3.8×10^{18} (ultraviolet)

9. 1.2×10^{15} (visible light)

10. 9.8×10^{12} (infrared)

11. 1×10^{11} (microwaves)

12. 9×10^8 (radar)

13. 92 mHz (92 million hertz) (FM radio station)

14. 660 kHz (660,000 hertz) (AM radio station)

Calculate the wavelength of sounds of the frequencies in Exercises 15 to 19.

15. 256 hertz (middle C)

16. 128 hertz (C below middle C)

17. 512 hertz (C above middle C)

18. 2000 hertz (piccolo range sound)

19. 80 hertz (low bass voice)

★20. Suppose you have two waves with the same frequency, amplitude, and velocity but slightly different initial phases. In particular, assume wave 1 is given by the function

$$y_1 = A\sin(kx - \omega t)$$

and wave 2 is given by the function

$$y_2 = A\sin(kx - \omega t - B)$$

What will be the total wave $(y_1 + y_2)$? Will the amplitude of the total wave be greater than A or less than A?

★21. Suppose that you have two waves with the same amplitude and velocity but slightly different frequencies. Let wave 1 be

$$y_1 = A\sin(k_1 x - \omega_1 t)$$

and let wave 2 be

$$y_2 = A\sin(k_2 x - \omega_2 t)$$

Calculate an approximate expression for the total wave $(y_1 + y_2)$. Assume that $\omega_2 - \omega_1$ is much smaller than ω_1, and assume $k_2 - k_1$ is much smaller than k_1.

☐Sketch graphs of the curves for Exercises 22–26. Use the program from Chapter 8, Exercise 19.

★22. $y = \sin x + \sin(2x) + \sin(3x)$

★23. $y = \sin x + 0.5\sin(2x) + 0.25\sin(4x)$
$\quad + 0.125\sin(8x) + 0.062\sin(16x)$

★24. $y = 0.062\sin x + 0.125\sin(2x) + 0.25\sin(4x)$
$\quad + 0.5\sin(8x) + \sin(16x)$

★25. $y = 1/4\sin x + 1/3\sin(2x) + 1/2\sin(3x) + \sin(4x)$
$\quad + 1/2\sin(6x) + 1/3\sin(7x) + 1/4\sin 8x$

★26. $y = \sin x + 1/3\cos(3x) + 1/5\sin(5x)$
$\quad + 1/7\cos(7x) + 1/9\sin(9x)$

★27. Add more terms to the series in Exercise 26, following the same pattern, and see what the graph looks like.

★28. Suppose an organ pipe of length L is open at one end. Standing waves will form in the pipe when the organ is played. Let $y = A\sin kx$ describe the wave pattern when the wave is at its maximum. At $x = 0$, $y = 0$; at $x = L$, $y = A$. Find the allowable values for the wavelength.

10
Inverse Trigonometric Functions

"There's no way we'll be able to send a message explaining Builder's design to Peaceful Bay in time!" Recordis moaned.

"We will not give up!" the king said.

At that moment Pal came skipping by playing with his pet pigeons.

"Not now!" the professor said. "We do not have time to play with pigeons!"

"The pigeons are fast flyers!" the king suddenly realized. "Maybe one of them can reach Peaceful Bay in time!"

"But those pigeons are totally scatterbrained!" the professor said. "How will they know how to get there?"

Pal's Pet Pigeons to the Rescue

"The pigeons are smarter than they look," Recordis said. "They know that whenever Pal releases them from High Tower they are supposed to fly in a perfectly straight line in the direction they are pointed."

"Then this is a trigonometry problem!" Trigonometeris said excitedly. "All we need to do is figure out the proper direction and then point the pigeons in that direction!"

"We know that Peaceful Bay is 400 miles north of Capital City and 300 miles to the east," the king said. (See Figure 10-1.)

Figure 10-1

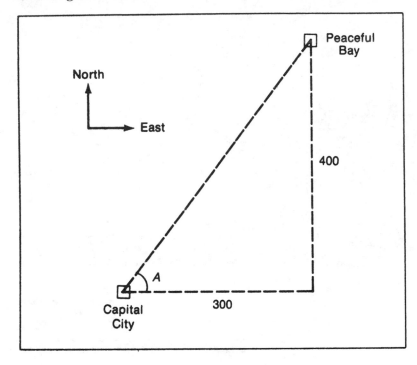

"All we need to do is set up a right triangle," Trigonometeris said. "Let's use the letter A to represent the angle between the ray pointing to Peaceful Bay and the ray pointing directly east. Then we can easily see that

$$\tan A = \frac{400}{300}$$

$$= \frac{4}{3}$$

"From this information we can clearly see that A is equal to. . . ." Suddenly Trigonometeris stopped and broke into a cold, desperate sweat.

"What's the problem?" the professor asked.

We all stared at the equation $\tan A = \frac{4}{3}$ and tried to figure out the value of A. Then we all realized the problem.

"If we know the value of A, then we can easily use our calculator to find the value of $\tan A$," Recordis said. "However, if we don't know the value of A, but we do know the value of $\tan A$, then there is no way to find the value of A. We can't use the calculator backward."

"Why not?" the professor asked excitedly. "All we need to do is work the tangent function backward! We faced this same problem many times before while we were working on algebra. We found many times that it was useful to develop an *inverse* function. An inverse function does the exact opposite of the original function. For example, the common logarithm function $y = \log x$ is the inverse function for the exponential function $x = 10^y$." She showed some examples:

$$2 = \log 100 \qquad\qquad 100 = 10^2$$
$$5 = \log 100{,}000 \qquad 100{,}000 = 10^5$$
$$1 = \log 10 \qquad\qquad 10 = 10^1$$
$$0.3010 = \log 2 \qquad\qquad 2 = 10^{0.3010}$$

"I know another example of a function and its inverse function," the king said. "If we have the function $y = x^3$, then the inverse function is $x = \sqrt[3]{y}$. For example,

$$8 = 2^3 \qquad 2 = \sqrt[3]{8}$$
$$27 = 3^3 \qquad 3 = \sqrt[3]{27}$$
$$64 = 4^3 \qquad 4 = \sqrt[3]{64}$$

"I know another example," the professor said. "Suppose $R = f(D) = \pi D/180$ is a function that converts an angle measured in degrees (D) into the same angle measured in radians. Then the inverse function is the function $D = g(R) = 180R/\pi$, which converts an angle measured in radians into the same angle measured in degrees."

"We know a lot of inverse functions," Recordis said, "but we don't have the faintest clue what function might be the inverse function for the tangent function." Trigonometeris, looking as if we wanted to crawl in a hole to hide his embarrassment, said nothing.

"Let's give a name to the inverse function first," the professor suggested. "Then we'll worry about how to calculate it later." She suggested the name angletan, but Recordis thought this name was too long. The professor came up with a new idea. Since angles reminded her of arcs of circles, she suggested the name *arctan*. We decided to accept that definition. The inverse function for the tangent function is

If $t = \tan A$, then $A = \arctan t$.

"We can also use the name *arcsin* to represent the inverse function for the sine function, and the name *arccos* to represent the inverse function for the cosine function," the professor continued, pleased that her idea had turned out to be so versatile.

> The inverse trigonometric functions are as follows:
>
> If $s = \sin A$, then $A = \text{arcsin } s$.
>
> If $c = \cos A$, then $A = \text{arccos } c$.
>
> If $t = \tan A$, then $A = \text{arctan } t$.
>
> We rewrote these relations in another way:
>
> $$\sin [\text{arcsin } (s)] = s$$
> $$\cos [\text{arccos } (c)] = c$$
> $$\tan [\text{arctan } (t)] = t$$

The results for inverse trigonometric functions may be expressed in either degrees or radians, depending upon which is most convenient.

(Inverse trigonometric functions can also be represented by another notation:

$$\text{arcsin } s = \sin^{-1} s$$
$$\text{arccos } c = \cos^{-1} c$$
$$\text{arctan } t = \tan^{-1} t$$

The −1 above each function stands for inverse function. However, if you use this notation you must be careful that you do not confuse the −1 used to represent inverse function with a −1 used as an exponent. Your calculator may use this notation on the keys representing the inverse trigonometric functions.)

"But we still don't know how to calculate values for any of these inverse functions," Recordis pointed out.

"We might turn out to be incredibly lucky and know the values already," Trigonometeris said hopefully. "For example, suppose we need to calculate arctan (1). That means we need to find an angle whose tangent is equal to 1, and I happen to know that $\tan(45°) = \tan(\pi/4) = 1$. Therefore, arctan(1) = 45°, or $\pi/4$ rad."

Trigonometer is made a list of all the values he knew by memory. It was an impressive list, although he had not yet succeeded in his original goal of memorizing the entire trigonometric table.

$$\arcsin 0 = 0 \qquad \text{since} \quad \sin 0 = 0$$

$$\arcsin \frac{1}{2} = \frac{\pi}{6} \, (30°) \qquad \text{since} \quad \sin \frac{\pi}{6} = \frac{1}{2}$$

$$\arcsin \frac{1}{\sqrt{2}} = \frac{\pi}{4} \, (45°) \qquad \text{since} \quad \sin \frac{\pi}{4} = \frac{1}{\sqrt{2}}$$

$$\arcsin \frac{\sqrt{3}}{2} = \frac{\pi}{3} \, (60°) \qquad \text{since} \quad \sin \frac{\pi}{3} = \frac{\sqrt{3}}{2}$$

$$\arcsin 1 = \frac{\pi}{2} \, (90°) \qquad \text{since} \quad \sin \frac{\pi}{2} = 1$$

$$\arccos 0 = \frac{\pi}{2} \, (90°) \qquad \text{since} \quad \cos \frac{\pi}{2} = 0$$

$$\arccos \frac{1}{2} = \frac{\pi}{3} \, (60°) \qquad \text{since} \quad \cos \frac{\pi}{3} = \frac{1}{2}$$

$$\arccos \frac{1}{\sqrt{2}} = \frac{\pi}{4} \, (45°) \qquad \text{since} \quad \cos \frac{\pi}{4} = \frac{1}{\sqrt{2}}$$

$$\arccos \frac{\sqrt{3}}{2} = \frac{\pi}{6} \, (30°) \qquad \text{since} \quad \cos \frac{\pi}{6} = \frac{\sqrt{3}}{2}$$

$$\arccos 1 = 0 \qquad \text{since} \quad \cos 0 = 1$$

$$\arctan 0 = 0 \qquad \text{since} \quad \tan 0 = 0$$

$$\arctan \frac{1}{\sqrt{3}} = \frac{\pi}{6} \, (30°) \qquad \text{since} \quad \tan \frac{\pi}{6} = \frac{1}{\sqrt{3}}$$

$$\arctan 1 = \frac{\pi}{4} \, (45°) \qquad \text{since} \quad \tan \frac{\pi}{4} = 1$$

$$\arctan \sqrt{3} = \frac{\pi}{3} \, (60°) \qquad \text{since} \quad \tan \frac{\pi}{3} = \sqrt{3}$$

"Those results are all obvious!" Recordis pointed out. "However, there is no obvious result for arctan $\frac{4}{3}$ = arctan 1.3333."

We asked Builder to work on this problem. Before long he presented us with a new, improved calculator that now had an arctan button. We entered the number 1.3333 and pressed the arctan button, and the computer displayed the result: 53.13 degrees. To verify that this result was correct, we entered the number 53.13 into the calculator and then pressed the tan button. Just as we expected, the result was 1.3333 (when rounded to four decimal places).

"The problem is solved!" the king said excitedly. "We must direct the pigeons to fly at an angle 53.13° north of due east." (See Figure 10-2.)

Figure 10-2

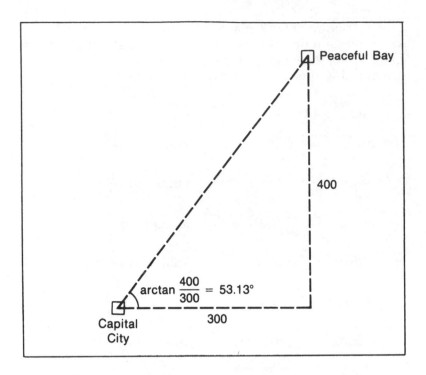

We quickly attached a little capsule containing Builder's design to one of the pigeons. Then Pal took them to the top of the tower and pointed them in the right direction. He released them, and they all started flying along a perfectly straight course in a direction 53.13° north of due east.

We all waited nervously. There was nothing more we could do now to save Peaceful Bay. To give us something to do while we waited, the professor suggested that we investigate more properties of the inverse trigonometric functions. "Let's find the domain and range of the functions."

"The domain of the arctan function consists of all real numbers," Trigonometeris said confidently. "Since the value of $\tan A$ can be any number from minus infinity to plus infinity, it follows that $\arctan t$ is defined for any value of t. The range of the function must be from 0 to 2π, since we know that the value of $\arctan t$ cannot be greater than 2π, since...." He stopped short.

"We have done something horribly wrong!" Recordis screamed. "How *do* we know that the value of the arctan function is never greater than 2π? For example, suppose we are looking for $z = \arctan 0$. In this case z could have the value 0, or it could have the value 2π, or the value 4π, or 8π, and so on...."

"This means that the arctan function isn't even a true function!" the professor said in shock. "We know that a function must always specify one unique value of the dependent variable for every value of the independent variable."

We puzzled over this problem for a long time.

"I don't think this will be a big problem," Trigonometeris said finally. "Normally, I am sure we will only be interested in the most convenient values. For example, we know arctan 1 could be equal to either $\pi/4$ or $2936\pi + \pi/4$, but normally it will be most natural to use the value arctan 1 = $\pi/4$."

"Let's specify *principal values* for each of the inverse trigonometric functions," the professor said. "For example, we know that arctan(0) = 0, 2π, 4π, and so on. But we can say that normally we will use the expression arctan 0 to represent the principal value 0."

We decided that the principal values of the arctan function would be between $-\pi/2$ and $\pi/2$. In other words, the expression arctan t would mean the value of A between $-\pi/2$ and $\pi/2$ such that tan $A = t$. For example, arctan $\sqrt{3}$ means $\pi/3$ instead of $32\pi + \pi/3$, and arctan 1 means $\pi/4$ instead of $2\pi + \pi/4$.

"Once we uniquely specify which value we wish to use, the arctan function becomes a legitimate function," Trigonometeris said with relief.

We also decided on principal values for the arcsin and the arccos function. We decided that the principal values of the arcsin function would also be between $-\pi/2$ and $\pi/2$. For example,

$$\arcsin\frac{1}{2} = \frac{\pi}{6} = 30°$$

$$\arcsin\frac{-1}{2} = \frac{-\pi}{6} = -30°$$

$$\arcsin\frac{\sqrt{3}}{2} = \frac{\pi}{3} = 60°$$

$$\arcsin\frac{-\sqrt{3}}{2} = \frac{-\pi}{3} = -60°$$

Recordis suggested that the principal values of the arccos function should also be between $-\pi/2$ and $\pi/2$. However, we realized we would have a problem if we used that definition, because then the value of arccos (x) would not be uniquely defined if x were positive (Would arccos ($1/\sqrt{2}$) be $\pi/4$ or $-\pi/4$?) and because the value of arccos x would not be within this range at all if x was negative. Therefore, we decided that the principal values of arccos x would be between 0 and π. For example,

$$\arccos \frac{1}{2} = \frac{\pi}{3} = 60°$$

$$\arccos \frac{-1}{2} = \frac{2\pi}{3} = 120°$$

$$\arccos \frac{\sqrt{3}}{2} = \frac{\pi}{6} = 30°$$

$$\arccos \frac{-\sqrt{3}}{2} = \frac{5\pi}{6} = 150°$$

"Let's make a graph of the inverse trigonometric functions," Trigonometeris said excitedly.

"Not another graph!" Recordis moaned.

Graphs of Inverse Trigonometric Functions

"This will be no problem," the professor said. "Once you have drawn the graph of a function, it is easy to draw the graph of its inverse. You merely need to interchange the x and y axes, which you can do by drawing the graph on a transparent sheet, then turning it over and rotating it 90°."

The graph of the arcsin function looked the same as the graph of the sine function, except that it had been turned on its side. (See Figure 10-3.)

Figure 10-3

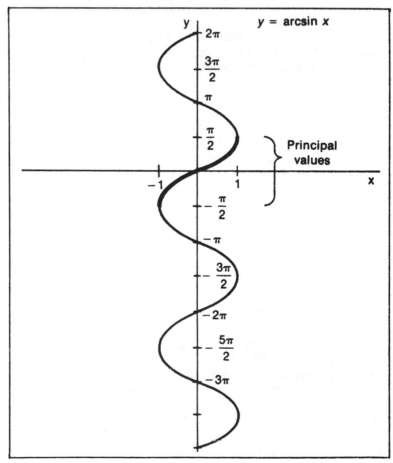

The graph of the arccos function looked the same as the graph of the arcsin function, except that it had been shifted down slightly. (See Figure 10-4.)

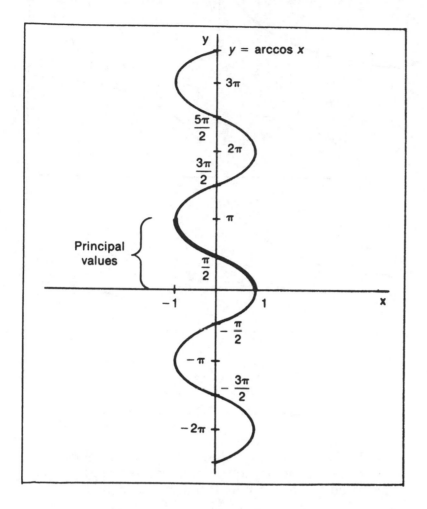

Figure 10-4

The graph of the arctan function consisted of a bunch of disconnected curves that looked like "esses" bent out of shape. (See Figure 10-5.)

Trigonometeris decided that for completeness we could use the name *arcsec* to represent the inverse of the secant function, the name *arccsc* to represent the inverse of the cosecant function, and the name *arcctn* to represent the inverse of the cotangent function. (The exercises provide more practice in dealing with the inverse trigonometric functions.)

We concluded our investigations for the day, but we still had to wait until we heard from Peaceful Bay. Finally, as evening approached, Pal's pigeons returned home. We quickly read the attached note, which told us that the gremlin's plot had been foiled and the people were safe!

Figure 10-5

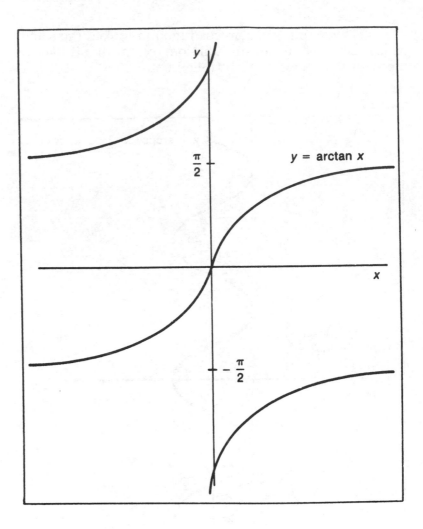

$y = \arctan x$

Exercises

In Exercises 1 to 7, find the principal values for the inverse trigonometric function expression. (Give an exact value when possible. If not, use a calculator.)

1. $\arcsin(-\frac{1}{2})$
2. $\arccos[(\sqrt{5}+1)/4]$
3. $\arctan(-1)$
4. $\arcsin 0.6$

5. $\arcsin 0.4$
6. $\arcsin(-0.3)$
7. $\arctan(5/12)$

In Exercises 8 to 15, you are given the lengths of the three sides of a triangle. Calculate the three angles.

8. 5, 6, 7
9. 12, 5, 13
10. 16, 16, 20
11. 19, 11, 29

12. 101, 101, 200
13. 4.35, 8.64, 5.72
14. 8.70, 17.28, 11.44
15. 14, 18, 22

In Exercises 16 to 24 you are given the length of two sides of a triangle and the size of the angle between those two sides. Calculate the length of the third side and the size of the other two angles.

16.	10, 15, 80°	21.	4.2, 11.8, 100°
17.	10, 15, 90°	22.	116, 120, 75°
18.	10, 15, 100°	23.	55, 32, 35°
19.	10, 15, 150°	24.	18, 20, 65°
20.	2.34, 6.18, 30°		

In Exercises 25 to 35, you are given the wind speed w, the air speed of a plane v, the plane's direction relative to the air A, and the wind's direction B. (See Chapter 7, Exercise 25.) Calculate the plane's groundspeed and direction relative to the ground.

	w	v	A	B
25.	25	500	15°	120°
26.	25	500	15°	250°
27.	25	500	15°	330°
28.	10	450	70°	10°
29.	10	450	70°	50°
30.	10	450	70°	85°
31.	10	450	70°	170°
32.	8	600	0°	40°
33.	8	600	0°	80°
34.	8	600	0°	110°
35.	8	600	0°	150°

Evaluate the trigonometric expression in Exercises 36 to 42.

36. $\tan(\arcsin\frac{3}{5})$

37. $\sin(\arctan\frac{12}{5})$

38. $\sin(\arccos\frac{8}{10})$

39. $\cos[\arcsin(-\frac{4}{5}) + \arccos\frac{12}{13}]$

40. $\sin(\arccos a)$

41. $\sin\left(\arccos\sqrt{1-a^2}\right)$

42. $\sin[\arctan(bx/a)]$

A *trigonometric equation* is an equation involving trigonometric functions. To solve the equation, you must find the values of x that make the equation true. For example, the equation $\sin x = \cos x$ has two solutions: $\pi/4$ and $5\pi/4$. Of course, any trigonometric equation that has at least one solution will also have an infinite number of solutions, but we will only be interested in the solutions that are between 0 and 2π. To solve these equations, make use of the trigonometric identities in Chapter 6. (Remember that an identity is a special type of equation that is true for all permissible values of the unknown.) In Exercises 43 to 63, solve the equations for x and use a graphing calculator or spreadsheet to illustrate the solutions.

43. $\sin x = \tan x$

44. $2\cos x - 1 = 0$

45. $\cos^2 x \sin^2 x = 0.8$

46. $\sin x + \cos x = 1$

★47. $\sin x + \cos x = \frac{1}{2}$

★48. $\sin^3 x = (1 - \cos 2x)/4$

★49. $16\sin^2 x - 16\sin^4 x = 3$

★50. $\sin^2 x + [(\sqrt{3} - 1)/2] = \sqrt{3}/4$

51. $\sin^2 x - (1/\sqrt{2})\sin x = 0$

52. $\tan^2 x = 3$

53. $2\tan^3 x + \tan x = 0$

★54. $4\sin x + 3\cos x = 2$

★55. $\arcsin 2x + \arcsin x = \pi/2$

56. $2\operatorname{ctn} x \cos x = \operatorname{ctn} x$

57. $[2\sin(2x) + 1](2\cos x + \sqrt{3}) = 0$

★58. $2\operatorname{ctn} x - \tan x - 1 = 0$

59. $\cos 2x = \sin x$

60. $\tan x \cos x - \cos x = 0$

61. $2\cos^2 x - 3\sin x = 3$

62. $2\sin^2 x - \sqrt{3}\sin x = 0$

★63. $3\cos^2 x + 5\cos x - 2 = 0$

★64. Solve this system of equations for A and B:

$$3\sin A + \cos B = 1$$
$$\sin A - \cos B = 1$$

★65. Show

$$\arcsin\frac{4}{5} = \pi - 2\arctan 2$$

66. Suppose you are given the coordinates of a point (x, y) and you would like to calculate the angle between the x axis and the line connecting the origin to that point. You will use the arctan function, but how will you make sure that the resulting angle is in the correct quadrant?

☐67. Suppose your programming language does not have the arcsin or the arccos functions, but it does have the arctan function. Write a program that you may use to calculate arcsin and arccos by using arctan.

11
Polar Coordinates

The next day we were still rejoicing about having saved the kingdom. Recordis explained how Pal had been able to make sure that the pigeons would stop when they reached Peaceful Bay. "Pal can control the distance the pigeons will fly by regulating the amount of scientifically designed birdseed he feeds them. For example, if he wants them to fly 50 miles he feeds them twice as much birdseed as when he wants them to fly 25 miles."

The king had a brilliant idea. "We should use the pigeons to provide a regular messenger service," he said. "That way we will be able to send messages all over the kingdom!"

"The Royal Map marks the location of every town in the kingdom," Recordis said. "We know how far north and how far east of Capital City every town is."

"Our system for identifying the locations of the towns is like a rectangular coordinate system," the professor said. "We have drawn a y axis that points north and an x axis that point east." (See Figure 11-1.)

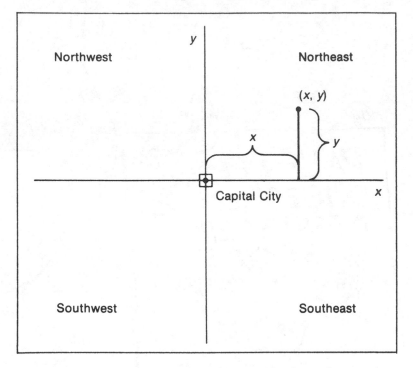

"Capital City is at the origin, which is the point $(0, 0)$. If a town is northeast of Capital City, then both the x coordinate and the y coordinate are positive. If a town is northwest of Capital City, then the y coordinate is positive and the x coordinate is negative. If the town is southwest of Capital City, then both the x coordinate and the y coordinate are negative, and if the town is southeast of Capital City then the x coordinate is positive and the y coordinate is negative." (A rectangular coordinate system is also called a Cartesian coordinate system, after René Descartes.)

"It doesn't do any good to tell the pigeons the rectangular coordinates of a town!" Recordis pointed out. "The pigeons can only find the town if they know the direction to the town."

"That sets the stage for my latest idea," the professor said. "I know of a totally new way to keep track of the locations of the towns in the kingdom. The rectangular xy system is fine for some purposes. However, for giving directions to the pigeons, we can use a new system. We can identify the location of each town with two numbers: the distance from that town to Capital City, and the direction you need to travel to get to that town. We will measure directions like this: we will say that east is 0°, north is 90°, west is 180°, and south is 270°." (See Figure 11-2.)

Figure 11-2

```
                    90° │ North
                        │
                        │
                        │
                        │
                        │
 180°                   │                    0°
─────────────────────────────────────────────
 West                   │                    East
                        │
                        │
                        │
                        │
                    270° │ South
```

"That is just the information we need to give to the pigeons!" the king said excitedly. "If we point them in the right direction and tell them the distance to the town, they will be able to find it."

We decided to call the new method of locating points the system of *polar coordinates*.

Polar Coordinates

Polar Coordinates

Any point in a plane can be identified by two numbers under the polar coordinate system. First, pick a point to represent the origin. Then, pick a direction to represent the 0° direction. We will always draw the 0° direction as pointing directly right from the origin. Then, any point in the plane can be identified by two coordinates called r and θ. (The symbol θ is a Greek letter called theta. The letter θ is another one of the professor's favorite Greek letters.)

r = distance from the origin to the point

θ = angle between the 0° line and the line drawn from the origin to the point (Figure 11-3)

Figure 11-3

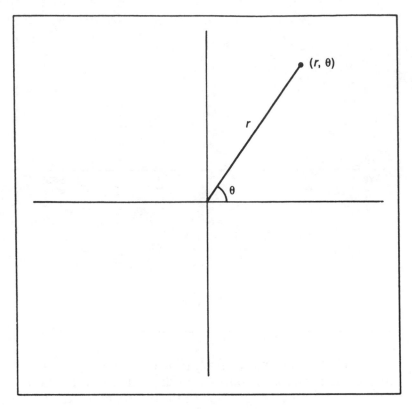

"This still leaves us with one slight problem," Recordis said. "We know the rectangular coordinates for each town. However, we don't know the polar coordinates for the towns."

"I know how we can convert the rectangular coordinates (x, y) to the polar coordinates (r, θ)," Trigonometeris said. "We can calculate r from the Pythagorean theorem:

$$r = \sqrt{x^2 + y^2}$$

And, since $\tan\theta = y/x$, we can calculate the value of θ by using the arctan function

$$\theta = \arctan\frac{y}{x}$$

We calculated some examples of conversions:

x	y	r	θ
15	15	21.2	45°
50	86.6	100	60°
0	10	10	90°
−17	17	24.04	135°
−111	−45	119.8	202°
1	−0.2679	1.035	−15°

"What if we need to do the reverse calculation?" Recordis demanded. "Suppose we know the polar coordinates of a town but we need to know the rectangular coordinates?"

Trigonometeris explained that we could use these formulas

$$x = r\cos\theta$$
$$y = r\sin\theta$$

Here are some examples.

r	θ	x	y
12	45°	8.49	8.49
16	30°	13.86	8.00
96	60°	48.00	83.14
4.34	123.5°	−2.40	3.62
0.075	218.9°	−0.058	−0.047
19	300°	9.5	−16.45

The king issued a proclamation.

To convert rectangular coordinates (x, y) to polar coordinates (r, θ),

$$r = \sqrt{x^2 + y^2}$$
$$\theta = \arctan\frac{y}{x}$$

To convert polar coordinates (r, θ) to rectangular coordinates (x, y),

$$x = r\cos\theta$$
$$y = r\sin\theta$$

(*Note*: You will need to use the rule described in Chapter 10, Exercise 66, to make sure that your result for θ is in the correct quadrant.)

Recordis put himself to work on the task of converting the rectangular coordinates of each town in the kingdom into polar coordinates. Builder started work on a sign reading, "Pal's Pet Pigeon Messenger Service—Fast Service to Any Town in Carmorra!" The professor wanted to investigate some more properties of the polar coordinate system.

"We have found it very useful to write equations to represent different shapes," she said thoughtfully. "We have written equations containing x and y and then drawn the graphs of the equations in rectangular coordinates. For example, we found that a circle with center at the origin and radius a could be represented by the equation

$$x^2 + y^2 = a^2$$

I wonder if we can draw figures in polar coordinates by finding equations containing r and θ."

"That will be easy," the king said. "All we need to do is start with an equation containing x and y. Then use the conversion equations

$$x = r\cos\theta$$

$$y = r\sin\theta$$

to convert the original xy equation into an equation containing r and θ."

Equations in Polar Coordinates

We tried to find the polar coordinate equation of a circle by substituting into the equation $x^2 + y^2 = a^2$:

$$(r\cos\theta^2) + (r\sin\theta^2) = a^2$$

$$r^2\cos^2\theta + r^2\sin^2\theta = a^2$$

$$r^2(\cos^2\theta + \sin^2\theta) = a^2$$

$$r^2 = a^2$$

$$r = a$$

"That equation is obvious!" Trigonometeris said. "We should have known that the polar coordinate equation of a circle would be $r = a$. That equation merely tells you to take all the points at a distance a from the origin, and we know that the definition of a circle is the set of all points a fixed distance from the center."

The professor decided to investigate what other kinds of curves we could draw in polar coordinates. Fortunately, Builder modified our graphing calculator so that it could draw curves in polar coordinates. You should find out how to create these graphs on your calculator. Also, the appendix shows how to create polar coordinate graphs with a computer spreadsheet. We found the equation for a horizontal line at a distance d away from the origin:

$$\frac{d}{r} = \sin\theta$$

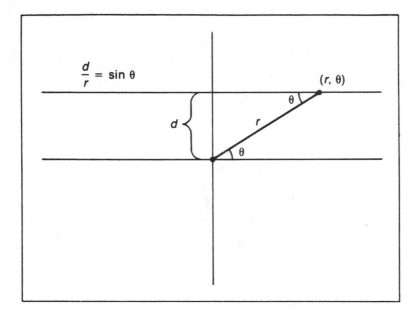

Figure 11-4

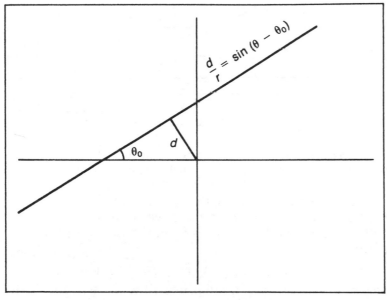

Figure 11-5

(See Figure 11-4.)

Then we found an equation for a line at a distance d away from the origin that was tilted at an angle θ_0:

$$\frac{d}{r} = \sin(\theta - \theta_0)$$

(See Figure 11-5.)

"Let's make up some equations and see what the graphs look like," Trigonometeris suggested. He suggested the equation

$$r = \theta$$

We found that the graph of this equation was an interesting spiral pattern (Figure 11-6). Next Trigonometeris suggested the equation

Figure 11-6

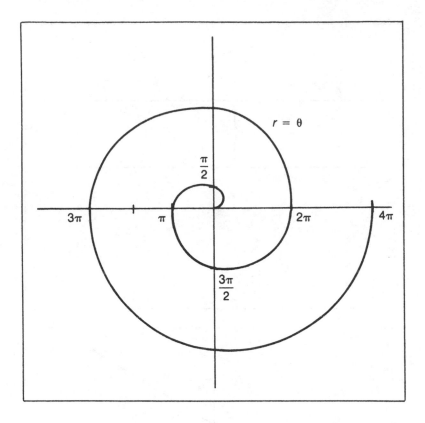

$$r = \cos\theta$$

We had drawn part of the graph when suddenly we ran into a problem. "When θ is greater than $\pi/2$, the value for $\cos\theta$ becomes negative," Recordis said. "We can't draw a point with a negative value for r. We know that a distance must always be positive, and r represents the distance from the origin to the point."

"I have an idea of what it means to have a negative value or r as a polar coordinate," the professor said. "We know that the point (r, θ) means the point at a distance r away from the origin in the direction given by the angle θ. So, logically, the point $(-r, \theta)$ should represent a point a distance r away from the origin in the opposite direction." (See Figure 11-7.)

Recordis, knowing that the professor usually got her way in such matters, decided not to protest.

We completed the graph of the equation $r = \cos\theta$ and found that it formed a circle. (See Figure 11-8.) This time the center of the circle was not at the origin.

"I bet I know how we can draw a curve with several loops," the professor guessed. She suggested we draw the graph of the curve $r = \sin 2\theta$. We made a table of values.

Figure 11-7

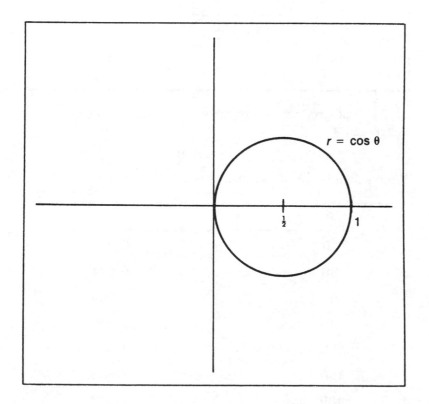

Figure 11-8

θ	r
0°	0
30°	0.866
60°	0.866
90°	0
120°	−0.866
150°	−0.866
180°	0
210°	0.866
240°	0.866
270°	0
300°	−0.866
330°	−0.866

Then we plotted the curve. (See Figure 11-9.)

Figure 11-9

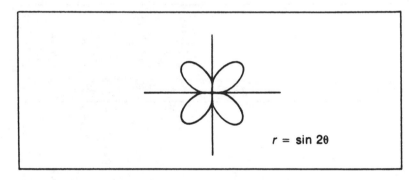

$r = \sin 2\theta$

"Let's try one more curve," Trigonometeris suggested.

$$r = 1 - \cos\theta$$

We made a table of values.

θ	r
0°	0
30°	0.134
60°	0.500
90°	1.000
120°	1.500
150°	1.866
180°	2.000

θ	r
210°	1.866
240°	1.500
270°	1.000
300°	0.500
330°	0.134

We started to plot the points. We stared in awe as the curve took shape. (See Figure 11-10.)

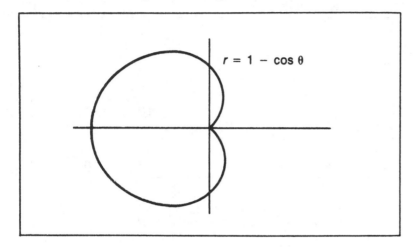

$r = 1 - \cos\theta$

Figure 11-10

"A Valentine's Day heart!" Recordis cried with delight. "I had always thought trigonometry was a cruel, uncaring subject, but I see now that sometimes trigonometry does have a heart." From that moment on Recordis began to like trigonometry a little bit. However, he never ceased to tease Trigonometeris about his favorite subject.

We were all exhausted from our ordeals. Now that the kingdom was safe we took a brief vacation from our investigation of trigonometry. The gremlin did not appear again for a long time. The pigeons provided a swift, economical communication system that linked the entire kingdom together. Many people found applications for trigonometry, including surveyors, navigators, physicists, musicians, engineers, and astronomers.

This brings to a conclusion the main part of our adventures. However, we did have four more trigonometry adventures. I have included these in case you feel that you have not yet satisfied your appetite for trigonometry. To appreciate the remaining adventures you will need to have a good understanding of such algebraic topics as complex numbers, conic sections, translations of coordinate axes, polynomials, and multiple linear equation systems. (If you are interested you may read the book *Algebra the Easy Way*, which tells how the people of Carmorra became acquainted with these and other algebra topics. If you would like to read about further adventures in the land of Carmorra, you may read the book *Calculus the Easy Way*. We found that a knowledge of trigonometry was very helpful during the course of our investigations of the subject of the calculus.)

Exercises

Represent each of the points in Exercises 1 to 12 in polar coordinates.

	x	y		x	y
1.	16	16	7.	6	−8
2.	7	26	8.	5	12
3.	−1	−2	9.	−7	−24
4.	−11	5	10.	11	−11
5.	4	−17	11.	14	7
6.	−3	4	12.	18	17

Represent each of the points in Exercises 13 to 24 in Cartesian (rectangular) coordinates.

	r	θ		r	θ
13.	10	90°	19.	18	150°
14.	5	0°	20.	45	23°
15.	117	270°	21.	16	7°
16.	39	180°	22.	26	23°
17.	15	45°	23.	10	−9°
18.	100	135°	24.	18	−11°

Sketch graphs of the polar coordinate equations in Exercises 25 to 33. Use a graphing calculator or a computer. Or you may find it helpful to obtain special polar coordinate graph paper, which you can either purchase or make copies of Figure 11-11.

25. $r = 2$

26. $r = 2\theta$

27. $r = \sin\theta$

28. $r = \sin^2 2\theta$

29. $r = \sin 3\theta$

30. $r = \sin 4\theta$

31. $r = 3(1 - \cos\theta)$

32. $r = 2 + \cos\theta$

33. $r\cos(\theta + \pi/6) = 3$

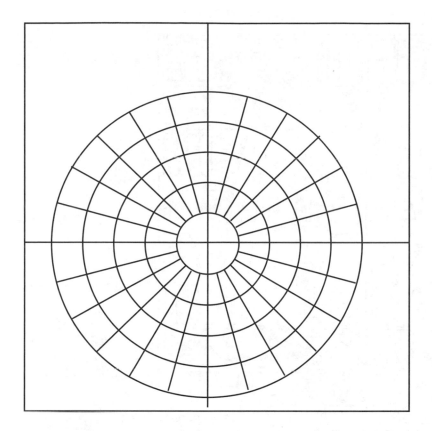

Figure 11-11

★34.   Change the equation $r^2 = \sec 2\theta$ to Cartesian coordinates and sketch its graph.

★35.   Find all the points where these two curves intersect:

$$r = 4(1 + \cos\theta)$$

$$r(1 - \cos\theta) = 3$$

★36.   In Cartesian coordinates the equation of a line is usually given as $y = mx + b$, where m is the slope of the line and b is the y intercept of the line. Show that this form of the equation is equivalent to the equation $d/r = \sin(\theta - \theta_0)$ given in the chapter by finding expressions for m and b in terms of d and θ_0.

□37.   Write a program to draw graphs of curves in polar coordinates. Test it out on the example curves given in the chapter. Also test it out on this curve:

$$r = \sin(n\theta)$$

Test several values for n, both whole numbers and fractions, and see if you can describe how the value of n determines the pattern of the graph.

12
Complex Numbers

One day Trigonometeris was searching through the Royal Archives. Recordis was out for the day, so it was difficult to find anything. Trigonometeris was looking at documents that told of our discovery of algebra when he came across an interesting folder labeled "Complex Numbers." "What are these?" he asked the professor.

"That's a new type of number we invented," the professor said. "We started with the *real numbers.* Every real number corresponds to a point on a number line, and a real number can be represented as a decimal fraction that either terminates, repeats the same pattern, or continues endlessly without ever repeating a pattern. However, we found there was no real number equal to the square root of −1. In other words, the equation $x^2 = -1$ has no real-number solutions. So, we made up a new number, called i, such that $i^2 = -1$. Of course, i cannot be a real number, but we decided to make it up anyway to see how it behaved. The gremlin dared us to do this. We called it an imaginary number."

"Then what is a complex number?" Trigonometeris asked.

"A complex number is a number like

$$a + bi$$

"where a and b are both real numbers. We call a number of the form bi a *pure imaginery* number. A complex number is formed by adding a real number and a pure imaginary number, although I should warn you that addition in this sense is not exactly the same as the ordinary addition you use when adding together two real numbers. In the complex number $a + bi$, a is the *real part* and b is the *imaginary part*. However, we must be very careful we do not let Recordis hear us talking about complex numbers, because he suffers fainting spells at the mere mention of the phrase 'complex number.'"

The folder in the archives listed some properties.

To add two complex numbers,

$$(a_1 + b_1 i) + (a_2 + b_2 i) = (a_1 + a_2) + (b_1 + b_2)i$$

To subtract two complex numbers,

$$(a_1 + b_1 i) - (a_2 + b_2 i) = (a_1 - a_2) + (b_1 - b_2)i$$

To multiply two complex numbers, treat each complex number as a binomial (and remember that $i^2 = -1$):

$$(a_1 + b_1 i)(a_2 + b_2 i)$$
$$= a_1 a_2 + a_1 b_2 i + a_2 b_1 i + b_1 b_2 i^2$$
$$= (a_1 a_2 - b_1 b_2) + (a_1 b_2 + a_2 b_1)i$$

A complex number can be represented on a two-dimensional diagram. The horizontal axis is the *real axis* and the vertical axis is the *imaginary axis*. The number $a + bi$ is represented by a point drawn a units to the right of the origin and b units up. (See Figure 12-1.)

The *absolute value* of a complex number is the distance from the origin to the point representing that number. We will use r to represent the absolute value. Then,

$$r = \sqrt{a^2 + b^2}$$

Here are some examples of complex numbers. (See Figure 12-2. *Note:* Real numbers are a special type of complex number.)

Figure 12-1

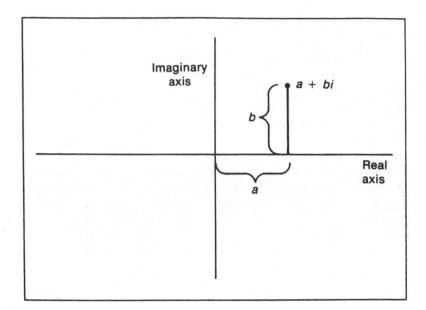

Figure 12-2

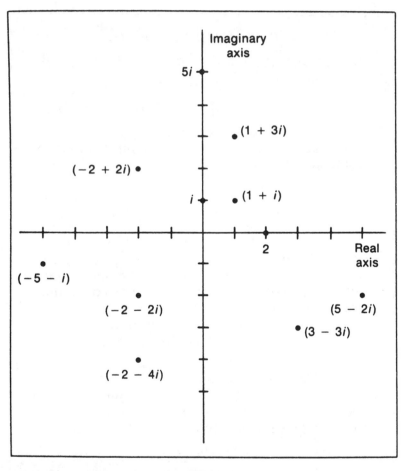

Trigonometeris stared at this diagram. "This method for representing complex numbers looks very much like a rectangular coordinate system," he said. "We found it was useful to convert rectangular coordinates into polar coordinates, so perhaps we can find a way of representing complex numbers using polar coordinates."

★**Polar Coordinate Form of Complex Numbers**

The professor became intrigued by the idea. "We know that a point in polar coordinates is represented by two quantities: the distance from the point to the origin, and the angle representing the direction you must travel to reach that point. It seems we should be able to represent a complex number by its absolute value and the angle between the real axis and the line connecting its point to the origin. In other words, a complex number is like a vector, or arrow: it has both length and direction."

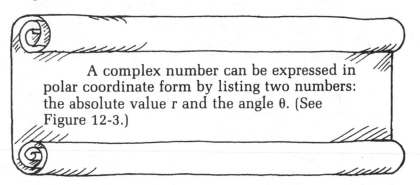

A complex number can be expressed in polar coordinate form by listing two numbers: the absolute value r and the angle θ. (See Figure 12-3.)

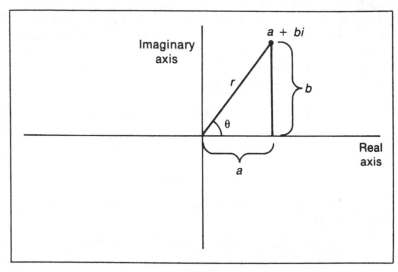

Figure 12-3

"We can use trigonometry to convert from the polar form to the regular form," Trigonometeris said.

$$a = r\cos\theta$$
$$b = r\sin\theta$$

"We can also convert the regular form to the polar form," the professor said.

$$r = \sqrt{a^2 + b^2}$$
$$\theta = \arctan\frac{b}{a}$$

We decided that normally we would write polar-form complex numbers like

$$r(\cos\theta + i\sin\theta)$$

"Let's make a list of some complex numbers expressed in both forms," Trigonometeris suggested.

$$5 = 5(\cos 0 + i \sin 0)$$

$$i = 1(\cos 90° + i \sin 90°)$$

$$7i = 7(\cos 90° + i \sin 90°)$$

$$1 + i = \sqrt{2}(\cos 45° + i \sin 45°)$$

$$1 - i = \sqrt{2}[\cos(-45°) + i \sin(-45°)]$$

$$3 + 3i = 3\sqrt{2}(\cos 45° + i \sin 45°)$$

$$1 + \sqrt{3}i = 2(\cos 60° + i \sin 60°)$$

$$3 - 4i = 5[\cos(-53.13°) + i \sin(-53.13°)]$$

$$-5 + 2i = \sqrt{29}(\cos 158.2° + i \sin 158.2°)$$

"This is all very interesting, but we have yet to find an advantage to writing complex numbers this way," the professor pointed out.

★Multiplying Complex Numbers

Trigonometeris was sure there must be some reason it would be more convenient to write complex numbers in polar notation. We found an addition rule for polar complex numbers, but it did not seem to be much of an improvement over the regular addition rule. (See Exercise 34.) However, some amazing results happened when we tried to multiply two complex numbers written in polar form:

$$[r_1(\cos \theta_1 + i \sin \theta_1)][r_2(\cos \theta_2 + i \sin \theta_2)]$$
$$= r_1 r_2[(\cos \theta_1 + i \sin \theta_1)(\cos \theta_2 + i \sin \theta_2)]$$
$$= r_1 r_2[\cos \theta_1 \cos \theta_2 - \sin \theta_1 \sin \theta_2$$
$$+ i(\sin \theta_1 \cos \theta_2 + \sin \theta_2 \cos \theta_1)]$$

"We can use the trigonometric addition rules!" Trigonometeris said.

$$[r_1(\cos \theta_1 + i \sin \theta_1)][r_2(\cos \theta_2 + i \sin \theta_2)]$$
$$= r_1 r_2[\cos(\theta_1 + \theta_2) + i \sin(\theta_1 + \theta_2)]$$

To multiply two complex numbers written in polar form,

$$[r_1(\cos \theta_1 + i \sin \theta_1)][r_2(\cos \theta_2 + i \sin \theta_2)]$$
$$= r_1 r_2[\cos(\theta_1 + \theta_2) + i \sin(\theta_1 + \theta_2)]$$

That is, to obtain the absolute value of the product, multiply the two absolute values.

To obtain the angle of the result, add the two angles.

Therefore, we wrote the final rule.

We worked some examples.

$[2(\cos 35° + i\sin 35°)][10(\cos 12° + i\sin 12°)]$
$= 20(\cos 47° + i\sin 47°)$

$[\sqrt{2}(\cos 45° + i\sin 45°)][\sqrt{2}(\cos 135° + i\sin 135°)]$
$= 2(\cos 180° + i\sin 180°)$
$= -2$

$[7(\cos 90° + i\sin 90°)][4(\cos 180° + i\sin 180°)]$
$= 28(\cos 270° + i\sin 270°)$

$(1 + i)(1 + \sqrt{3}i)$
$= [\sqrt{2}(\cos 45° + i\sin 45°)][2(\cos 60° + i\sin 60°)]$
$= 2\sqrt{2}(\cos 105° + i\sin 105°)$

We found that the rule even worked when multiplying a real number by a complex number. For example, suppose x is a positive real number. Then,

$$x = x(\cos 0 + i\sin 0)$$
$$x[r(\cos\theta + i\sin\theta)] = xr(\cos\theta + i\sin\theta)$$

"I see," the professor said. "The absolute value of the complex number is multiplied by x, but the angle remains unchanged when you multiply by a positive real number."

The professor also noticed an interesting effect if you multiplied a complex number by i:

$$i[r(\cos\theta + i\sin\theta)] = 1r[\cos(\theta + 90°) + i\sin(\theta + 90°)]$$

"Note that the absolute value of the number stays the same, but the angle has increased by 90°. This is the same as rotating the vector representing the number counterclockwise by 90°. It seems to me that multiplying by i is a signal to rotate by 90°."

We tried some simple examples to confirm that multiplying by i means to rotate the vector representing the number by 90 degrees:

Original number	After multiplying by i
10	$10i$
$10i$	$10i^2 = -10$
-10	$-10i$
$-10i$	$-10i^2 = 10$

"If you multiply by i twice, then you would rotate the vector by 180 degrees," Trigonometeris suggested the next logical step.

"Agreed," the professor said. "For example, if you start with 1 and multiply by i, you get i; if you multiply by i again, you get $i^2 = -1$, which is indeed the opposite direction (180 degrees) from 1." Suddenly her eyes widened. "Now I see why i is the square root of negative 1! Multiplying a complex number by -1 will shift its direction by 180 degrees—that is, it will point in the opposite direction. Multiplying a complex number by i shifts its direction by 90 degrees; and multiplying by i twice is exactly the same as multiplying by -1. In other words, $i^2 = -1$. It's too bad we used the term *imaginary number* for i because we didn't know any real meaning for that number when we first discovered it. But now we know what multiplying by i means on the graph of complex numbers."

"I see what will happen if you multiply by any complex number that has an absolute value of 1," the king noticed. "The absolute value of the original complex number will stay the same, but it will be rotated a certain amount." The king used $(\cos\theta_1 + i\sin\theta_1)$ to represent an arbitrary complex number with absolute value 1 and $r(\cos\theta_2 + i\sin\theta_2)$ to represent another complex number. Then he calculated

$$(\cos\theta_1 + i\sin\theta_1)r(\cos\theta_2 + i\sin\theta_2)$$
$$= r[\cos(\theta_1 + \theta_2) + i\sin(\theta_1 + \theta_2)]$$

★Powers of Complex Numbers

"We can now find powers for complex numbers!" the professor realized. "I remember that calculating powers of complex numbers written in regular form was very tedious, and since Recordis can't stand complex numbers, I ended up doing all the work."

To get the square of a complex number we found

$$[r(\cos\theta + i\sin\theta)]^2 = r^2(\cos 2\theta + i\sin 2\theta)$$

There's no reason to stop at 2:

$$[r(\cos\theta + i\sin\theta)]^3 = r^3(\cos 3\theta + i\sin 3\theta)$$

$$[r(\cos\theta + i\sin\theta)]^4 = r^4(\cos 4\theta + i\sin 4\theta)$$

$$[r(\cos\theta + i\sin\theta)]^5 = r^5(\cos 5\theta + i\sin 5\theta)$$

We found that, in general,

$$[r(\cos\theta + i\sin\theta)]^n = r^n(\cos n\theta + i\sin n\theta)$$

We worked some examples.

$$(1 + i)^6 = [\sqrt{2}(\cos 45° + i\sin 45°)]^6$$
$$= 8[\cos(6 \times 45°) + i\sin(6 \times 45°)]$$
$$= 8(\cos 270° + i\sin 270°)$$
$$5^3 = [5(\cos 0 + i\sin 0)]^3$$
$$= 125(\cos 0 + i\sin 0)$$

$$i^5 = [1(\cos 90° + i\sin 90°)]^5$$
$$= 1(\cos 450° + i\sin 450°)$$
$$= i$$
$$(1-i)^{10} = \{\sqrt{2}[\cos(-45°) + i\sin(-45°)]\}^{10}$$
$$= 32[\cos(-450°) + i\sin(-450°)]$$
$$= -32i$$
$$(3-4i)^2 = \{5[\cos(-53.13°) + i\sin(-53.13°)]\}^2$$
$$= 25[\cos(-106.26°) + i\sin(-106.26°)]$$

"If we can do powers, then we also should be able to do roots," the professor reasoned, "since taking a root of a number is the opposite of raising it to a power." She decided to look for the square root of i. (She knew that i was the square root of -1, but she had not yet been able to find a number that was the square root of i.)

"We'll write i in polar notation," she said.

$$i = \cos 90° + i\sin 90°$$

"To take the square root of a complex number in polar form, I bet we need to take the square root of the absolute value (which is 1 in this case) and divide the angle by 2:

$$\sqrt{i} = \cos 45° + i\sin 45°$$
$$= \frac{1}{\sqrt{2}} + i\frac{1}{\sqrt{2}}$$

"This answer has the added advantage of being right!" she said triumphantly after she checked to make sure that $(1/\sqrt{2} + i/\sqrt{2})$ squared was indeed equal to i. (See Exercise 47.)

"Is that the only square root?" the king asked. "We found that positive real numbers have two square roots, one positive and one negative. For example, $(-3)^2 = 9$ and $3^2 = 9$. (We used the radical symbol $\sqrt{}$ to always refer to the positive square root, but that doesn't mean we can ignore the negative square root.)"

We found that $-1/\sqrt{2} - i/\sqrt{2}$ was also a square root of i. The professor was puzzled about how this could be until she wrote the number in polar notation:

$$-\frac{1}{\sqrt{2}} - \frac{i}{\sqrt{2}} = \cos 225° + i\sin 225°$$

Then we squared that number by doubling its angle:

$$(\cos 225° + i\sin 225°)$$
$$= \cos(2 \times 225) + i\sin(2 \times 225)$$
$$= \cos 450° + i\sin 450°$$

"I see!" the professor said. "A 450° angle is coterminal with a 90° angle, so $(\cos 450° + i \sin 450°)$ is the same as $(\cos 90° + i \sin 90°)$, which is the same as i."

The professor thought a moment and made a shrewd guess. "I bet *every* complex number has two square roots." She thought a bit more. "I wonder if this means that every complex number also has three third roots, four fourth roots, five fifth roots, and so on."

We decided to look for cube roots of i:

$$i = \cos 90° + i \sin 90°$$

But we can also write that like

$$i = \cos 450° + i \sin 450°$$

$$i = \cos 810° + i \sin 810°$$

By using these three polar forms for i, we found three cube roots:

$$\sqrt[3]{i} = \cos 30° + i \sin 30°$$

$$\sqrt[3]{i} = \cos 150° + i \sin 150°$$

$$\sqrt[3]{i} = \cos 270° + i \sin 270°$$

We wrote a general rule.

Roots of Complex Numbers

A complex number has a total of n number of nth roots. For example, a complex number has one first root (itself), two square roots, three cube roots, four fourth roots, five fifth roots, and so on. Consider the complex number

$$r(\cos \theta_0 + i \sin \theta_0)$$

The n roots all have absolute value $r^{1/n}$. The n values of the angle θ can be found from the formula

$$\theta = \frac{360m + \theta_0}{n} \quad \text{degrees}$$

$$\theta = \frac{2\pi m + \theta_0}{n} \quad \text{radians}$$

The factor m takes on the values of all of the integers from 0 to $n - 1$.

We were so excited by the polar coordinate representation of complex numbers that we did not notice when Recordis walked into the Main Conference Room.

"Hi!" he said cheerfully. "What's new?"

The professor quickly covered the papers we had been writing on. "You had better not look."

Recordis looked puzzled. "Not even one little peek?" He caught a glimpse of a complex number on a corner of a page. "No! Not those numbers again!" he screamed and fainted.

Recordis never did overcome his fear of complex numbers, but we found that polar coordinate representation greatly enhanced our understanding of this unusual type of number.

Here are a few notes about advanced topics that are beyond the scope of this book but that you may encounter if you study calculus.

Notes to CHAPTER 12

- If you need to calculate the result when an exponent contains an imaginary number, use this formula:

$$a^{ix} = \cos(x \ln a) + i \sin(x \ln a)$$

where $\ln a$ represents the natural logarithm of a number. The natural logarithm function uses a base known as e, which is a special number used in calculus and is about 2.71828. If you are raising e to an imaginary power, the formula is even simpler:

$$e^{ix} = \cos x + i \sin x$$

We can use this formula to find a formula for $\cos x$ in terms of exponents:

$$\cos x = \frac{e^{ix} + e^{-ix}}{2}$$

- During the course of further study, you may encounter another type of cosine function called the *hyperbolic cosine*. This function is represented by the abbreviation *cosh* and is defined as follows:

$$\cosh x = \frac{e^{x} + e^{-x}}{2}$$

One application of this function is that it describes the shape of a flexible string or wire hanging between two supports. There also is a hyperbolic sine function:

$$\sinh x = \frac{e^x - e^{-x}}{2}$$

In Exercises 1 to 10, convert the complex numbers to polar form.

1. $3 - 4i$

2. $-12 - 5i$

3. $7.5i$

4. $-11.45i$

5. $2 + 2i$

6. $0.5 + 0.8666i$

7. $-11.4 + 34i$

8. $10 + 6i$

9. $-9 - 6i$

10. $7 + 24i$

In Exercises 11 to 20, convert each of the complex numbers expressed in polar form to regular form.

	r	θ
11.	5	53.13°
12.	13	22.62°
13.	25	−73.74°
14.	10	−36.87°
15.	1	−90°
16.	1	225°
17.	1	150°
18.	23.4	34.5°
19.	11.56	190.54°
20.	2.87	89.65°

21. The complex conjugate of a complex number $a + bi$ is equal to $a - bi$. In other words, to find the conjugate you reverse the sign of the imaginary part. Write a rule that tells how to find the conjugate of a complex number in polar form.

Perform the multiplications indicated in Exercises 22 to 31. Then convert the product and the two factors into regular form.

22. $[13(\cos 22.62° + i\sin 22.62°)]\, [5(\cos 53.13° + i\sin 53.13°)]$

23. $[10(\cos 36.87° + i\sin 36.87°)]$ $[25(\cos 16.26° + i\sin 16.26°)]$

24. $[50(\cos 83.74° + i\sin 83.74°)]$ $[10(\cos 143.13° + i\sin 143.13°)]$

25. $[2(\cos 60° + i\sin 60°)]$ $[2(\cos 150° + i\sin 150°)]$

26. $[2(\cos 240° + i\sin 240°)]$ $[2(\cos 300° + i\sin 300°)]$

27. $(\cos 45° + i\sin 45°)$ $(\cos 135° + i\sin 135°)$

28. $(\cos 45° + i\sin 45°)$ $(\cos 225° + i\sin 225°)$

29. $(\cos 45° + i\sin 45°)$ $(\cos 315° + i\sin 315°)$

30. $[16(\cos 59° + i\sin 59°)]$ $[97(\cos 26° + i\sin 26°)]$

31. $[99(\cos 13° + i\sin 13°)]$ $[33(\cos 23° + i\sin 23°)]$

★32. We found three cube roots of i in the chapter. Can you find additional third roots of i by writing i in any other coterminal forms?

★33. Can you state a general rule about the location on the real/imaginary diagram for the n nth roots of a complex number?

★34. Derive a rule that tells how to add together two complex numbers expressed in polar coordinate form.

★35. Derive a rule that tells how to divide two complex numbers expressed in polar coordinate form.

Calculate the powers of the complex numbers in Exercises 36 to 41.

36. $(6 + 8i)^4$

37. $(24 + 7i)^5$

38. $(1/\sqrt{2} + i/\sqrt{2})^3$

39. $(1/\sqrt{2} + i/\sqrt{2})^4$

40. $(1/\sqrt{2} + i/\sqrt{2})^{10}$

41. $(1/\sqrt{2} + i/\sqrt{2})^{63}$

Calculate the four fourth roots of the complex numbers in Exercises 42 to 46.

42. $1/\sqrt{2} + i/\sqrt{2}$

43. i

44. 1

45. $3 + 4i$

46. $16(\cos 80° + i\sin 80°)$

47.   Perform the following multiplications without using polar coordinates:

(a) $(1/\sqrt{2} + i/\sqrt{2})^2$

(b) $(-1/\sqrt{2} - i/\sqrt{2})^2$

(c) $(\sqrt{3}/2 + i/2)^3$

□48.   Write a program to calculate the n different nth roots of a complex number.

□49.   Write a program to calculate $(a + bi)^n$.

13

Coordinate Rotation and Conic Sections

The next week we received a visit from a tall, elegantly dressed gentleman. "This is Count Q, an old friend of the family," Recordis introduced him to us. The count joined us for lunch. He told us about some of his problems, and he explained how busy he was.

★ *The Peaceful Bay Town-Planning Problem*

"I spent my last vacation at the town of Peaceful Day, and while I was there, the people asked me to help with the Town Planning Commission. We are trying to make a map of the town. It has been very confusing. The center of the town is at the Town Triangle. We have measured the location of each point in the town by listing two numbers: the distance *north* from the Town Triangle to that point, and the distance *east* from the Town Triangle to that point."

"Ah!" the professor said. "Your system is just like a system of rectangular coordinates, with the *y* axis pointing north and the *x* axis pointing east."

"Correct," the count agreed. "Unfortunately, due to the unique geography of the town of Peaceful Bay,

187

that is not the best system to use in this case. The harbor line is inclined at an angle 20° north of east, and of course all the streets follow the harbor line. That is, they are either parallel to or perpendicular to the harbor." (See Figure 13-1.) "Therefore, the planning commission has decided to use a new system. The new system is a standard *xy* coordinate system. The only difference is that now the *x* axis is parallel to the harbor, and the *y* axis is perpendicular to the harbor. Note that the origin of the new system is still at the Town Triangle."

Figure 13-1

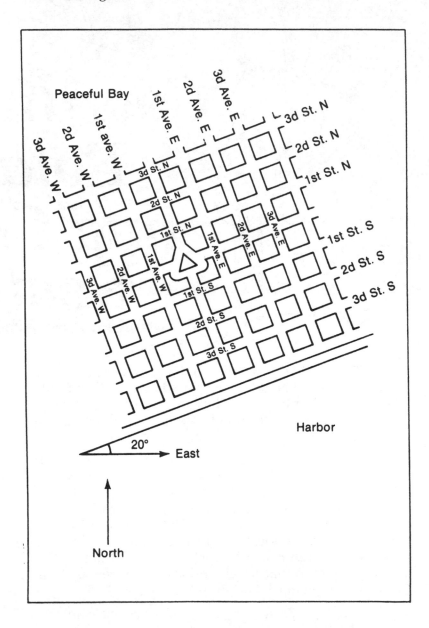

"How do you tell the difference between the coordinates measured in the old system and the coordinates measured in the new system?" Recordis asked.

"That was precisely the question I asked the other commission members," the count told us. "To keep from confusing the (x, y) coordinates in the new system with the (x, y) coordinates in the old system, we put a little mark ' (called a *prime* symbol) next to the letters. (That was my personal contribution to the system.) Therefore, we refer to the coordinates in the new system as (x', y'), as opposed to the (x, y) coordinates in the old system. For practical purposes the new system is more convenient because the coordinate axes match the street pattern. For example, we know that the point $(x' = 4, y' = 5)$ is at the corner of Fourth Avenue east and Fifth Street north." (See Figure 13-2.) "The coordinates of this same point in the old system are $(x = 2.048, y = 6.066)$."

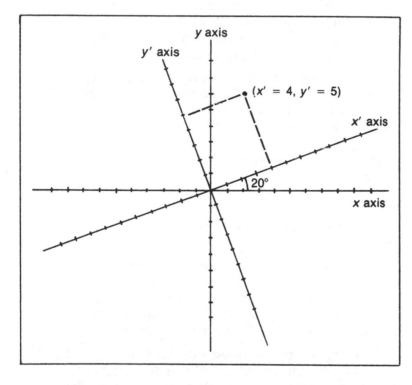

Figure 13-2

"The new system is much more convenient," Recordis agreed. "I like the numbers 4 and 5 much more than I like the numbers 2.048 and 6.066."

"However, we never intended that the new system would replace the old system," the count continued. "We merely intend to use it as a supplementary system. There are times when the old system based on north and east is more convenient. That leads us to our problem: we have measured the locations of many points using the old system. We would like a quick way to convert these old system coordinates (with no primes) into new system coordinates (with primes). Then we would not have to measure all the new system coordinates."

We were all silent. The count's problem sounded difficult. The professor decided to state the problem in more formal terms.

Coordinate Rotation Problem

Draw an x axis and a y axis. Any point in the plane can be identified by listing its two coordinates (x, y). Now, draw new coordinate axes called x' and y' (read "x prime" and "y prime"). These new axes are *rotated* by an angle θ_0 from the old x and y axes. Note that the origin of both coordinate systems is the same. Now, any point in the plane can also be identified by giving its two coordinates (x', y') in the new system.

The problem is, suppose you know the coordinates (x, y) of a point in the old system. How do you calculate the coordinates (x', y') of the same point in the new system?

★*Rotated Coordinate Systems*

"A very good restatement of the problem," the count agreed.

The count left us in the afternoon while we stayed in the Main Conference Room. Recordis told us that the count was quite rich and he would be sure to give us a generous present if we should be able to solve this problem for him. However, we had no success.

★*The New Improved Pigeon-Aiming System*

Later in the afternoon Builder stopped by the Main Conference Room. He told us that he had suggested some improvements for our pigeon-aiming system. "I'm sure you remember how the original system works," he said. Recordis nodded, pretending that he did. "We identify any location in the kingdom by specifying two numbers: r, the distance the pigeon must travel to reach the point, and θ, which is the direction to aim the pigeon. In the old system, we measured the angle as the angle north of east; in other words, the direction directly east is 0°, the direction directly north is 90°, west is 180°, and south is 270°. For most purposes, it is easiest to identify locations by the old system. However, for pigeon aiming, I find it is best to use a new system. In the new system, the 0° direction points to Distant Mountain." (See Figure 13-3.) "For example, the location of the town of Shady

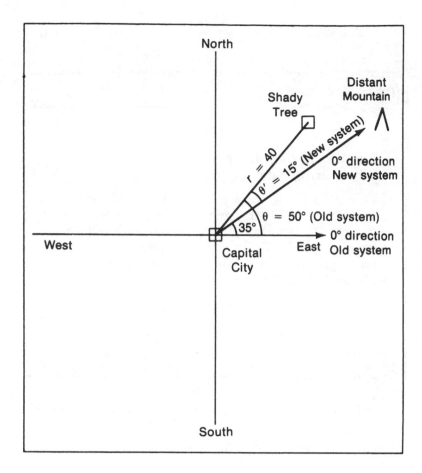

Figure 13-3

Tree is ($r = 40$, $\theta = 50°$) in the old system, but its location is ($r' = 40$, $\theta' = 15°$) in the new system."

"We've seen those little prime symbols before," Recordis said.

"I put primes after the coordinates in the new system so I don't confuse the new system coordinates with the old system coordinates," Builder explained.

"This is exactly the same as Count Q's coordinate rotation problem!" Trigonometeris cried. "Once again we are identifying points by using two different systems. In each system the origin is at the same location, but the axes have been rotated."

Recordis suddenly saw a big advantage to using polar coordinates. "It is very easy to convert from one coordinate system to a rotated coordinate system if you are using polar coordinates!" he exclaimed. "The distance from the origin to the point will be the same, no matter how you rotate the axes, so $r' = r$. If θ is the angle in the old system, θ' is the angle in the new system, and θ_0 is the angle of rotation, then

$$\theta' = \theta - \theta_0$$

★*Rotations in Polar Coordinates*

The king decreed:

**Rotation of Coordinates
When Using Polar Coordinates**

Consider an original polar coordinate system (r, θ). Then, create a new polar coordinate system by rotating the 0° direction by an angle θ_0. Call the coordinates in the new system (r', θ'). Then, to convert from the old system to the new system,

$$r' = r$$
$$\theta' = \theta - \theta_0$$

For example, if θ_0 (the angle of rotation) is 35°,

r	θ	r'	θ'
16	12.5	16	−22.5
100	38	100	3
45.4	175	45.4	140
11.4	240	11.4	205
19	345	19	310

★*Rotations in Rectangular Coordinates*

"Since we know how to handle a rotation when we are dealing with polar coordinates, and we know how to convert from polar coordinates to rectangular coordinates, we should be able to solve Count Q's problem involving rectangular coordinate rotation," the professor said.

Old (no prime) system	New (prime) system
$x = r\cos\theta$	$x' = r'\cos\theta'$
$y = r\sin\theta$	$y' = r'\sin\theta'$

"To convert from the old system to the new system,

$$r' = r$$
$$\theta' = \theta - \theta_o$$

"Let's substitute $\theta' = \theta - \theta_o$ and $r' = r$ into the equations $x' = r'\cos\theta'$ and $y' = r'\sin\theta'$," the king suggested.

$$x' = r\cos(\theta - \theta_o)$$
$$y' = r\sin(\theta - \theta_o)$$

"We can use the trigonometric subtraction rules," Trigonometeris said helpfully.

$$x' = r(\cos\theta\cos\theta_o + \sin\theta\sin\theta_o)$$
$$y' = r(\sin\theta\cos\theta_o - \sin\theta_o\cos\theta)$$

We rewrote those:

$$x' = r\cos\theta\cos\theta_o + r\sin\theta\sin\theta_o$$
$$y' = r\sin\theta\cos\theta_o - r\cos\theta\sin\theta_o$$

Using $x = r\cos\theta$ and $y = r\sin\theta$,

$$x' = x\cos\theta_o + y\sin\theta_o$$
$$y' = y\cos\theta_o - x\sin\theta_o$$

"That's the answer!" the professor exclaimed. "If we know the old coordinates (x, y) and the angle of rotation θ_o we can calculate the new coordinates (x', y')!"

"We should try some examples to make sure that it works," the king cautioned.

"I see one obvious example," Recordis said. "Suppose $\theta_o = 0$; in other words, suppose you are not rotating the axes at all. In that case,

$$x' = x\cos 0 + y\sin 0 = (x \times 1) + (y \times 0) = x$$
$$y' = y\cos 0 - x\sin 0 = (y \times 1) - (x \times 0) = y$$

The new coordinates are exactly the same as the old coordinates."

"I know of another example," the professor said. "Suppose you rotate the axes by 90°. Then, since $\cos 90° = 0$ and $\sin 90° = 1$, it follows that

$$x' = x\cos 90° + y\sin 90° = (x \times 0) + (y \times 1) = y$$
$$y' = y\cos 90° - x\sin 90° = (y \times 0) - (x \times 1) = -x$$

Count Q's problem involved a rotation of 20°, so we found

$$x' = x\cos 20° + y\sin 20°$$
$$y' = y\cos 20° - x\sin 20°$$
$$x' = 0.9397x + 0.342y$$
$$y' = 0.9397y - 0.342x$$

We wrote a general rule.

Coordinate Rotation

Suppose that a new coordinate system (x', y') is formed by rotating the axes of the old coordinate system (x, y) by an angle θ_0. If you know the old coordinates and would like to know the new coordinates, you may use the formula

$$x' = x \cos \theta_0 + y \sin \theta_0$$

$$y' = y \cos \theta_0 - x \sin \theta_0$$

If you know the new coordinates and would like to calculate the old coordinates, use the formula

$$x = x' \cos \theta_0 - y' \sin \theta_0$$

$$y = y' \cos \theta_0 + x' \sin \theta_0$$

(See Exercise 15 for a derivation of the reverse transformation formulas.)

We gave these results to Count Q, and he returned home after graciously promising us a generous gift.

The next morning the professor came running into the Main Conference Room triumphantly. "I have finally completed my detailed investigations of a quadratic equation with two variables!" she exclaimed.

"What?" Recordis asked blankly.

"You remember when we investigated the solution to a quadratic equation with one unknown, such as $ax^2 + bx + c = 0$? In general we found this type of equation usually has two solutions. Now, suppose we have an equation involving x values and y values, but no term has a higher degree than 2. For example,

$$x^2 + 5y^2 - 10x + 15y - 20 = 0$$

I call this type of equation a quadratic equation with two variables. And I have found that in general there will be many possible pairs of values (x, y) that are solutions to the equation. If you make a graph of the solution, then it will trace out a nice conic section,

★The Two-Unknown Quadratic Equation

such as a circle, an ellipse, a parabola, or a hyperbola."

"We should review conic sections," Recordis said. "There are a few minor details that seem to have slipped my mind."

The professor leafed through the archives and found a description of conic sections.

Conic Sections

The four curves—circles, ellipses, parabolas, and hyperbolas—are the *conic sections.*

1. Circles

A *circle* is the set of points in a plane that are all the same distance r from a fixed point called the center. The equation of a circle with center at the origin can be written

$$x^2 + y^2 = r^2$$

where r is the radius of the circle. (See Figure 13-4.)

★*Circles*

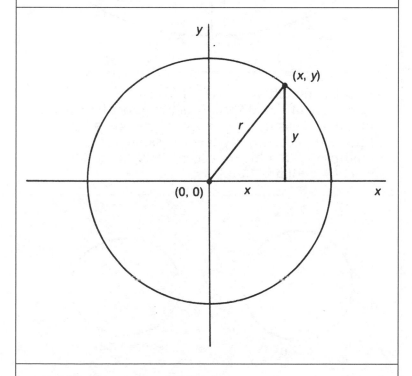

Figure 13-4

2. Ellipses

An *ellipse* is the set of points in a plane such that the sum of the distances to two fixed points is constant. The two fixed points are the *focal points.* The point halfway between the two focal points is called the center. The longest

★*Ellipses*

distance across the ellipse is the *major axis*; half this distance is the *semimajor axis*. The shortest distance across the ellipse is the *minor axis*; half this distance is the *semiminor axis*. The equation of an ellipse with its center at the origin, a semimajor axis of length a, and a semiminor axis of length b is

$$\frac{x^2}{a^2} + \frac{y^2}{b^2} = 1$$

(See Figure 13-5.)

Figure 13-5

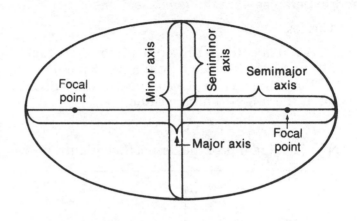

The quantity $e = \sqrt{a^2 - b^2}/a$ is the *eccentricity* of the ellipse. It is a number between 0 and 1 that measures the shape of the ellipse. An ellipse with eccentricity 0 is the same as a circle. An ellipse with a higher eccentricity has a flatter shape. (See Figure 13-6.)

Figure 13-6

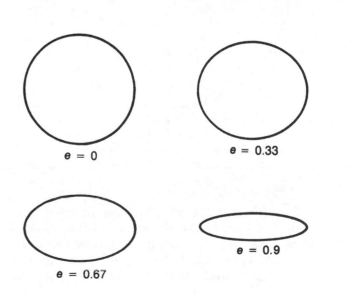

3. Parabolas

A *parabola* is the set of all points in a plane that are the same distance from a fixed line (the *directrix*) and a fixed point (the *focus*). The point on the parabola closest to the focus is the *vertex*. If the focus of a parabola is the point $(0, a)$ and the directrix is the line $y = -a$, then the vertex is at the point $(0, 0)$ and the equation of the parabola is

$$y = \frac{x^2}{4a}$$

(See Figure 13-7.)

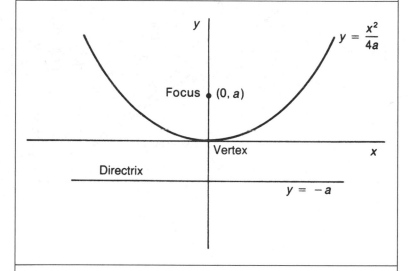

Figure 13-7

4. Hyperbolas

A *hyperbola* is the set of all points in a plane such that the difference between the distances to two fixed points is a constant. A hyperbola has two branches that are mirror images of each other. Each branch looks like a misshapen parabola. The general equation for a hyperbola with center at the origin is

$$\frac{x^2}{a^2} - \frac{y^2}{b^2} = 1$$

The meanings of a and b are shown in the diagram. The two diagonal lines are *asymptotes*. As x gets larger and larger, the positive branch of

Figure 13-8

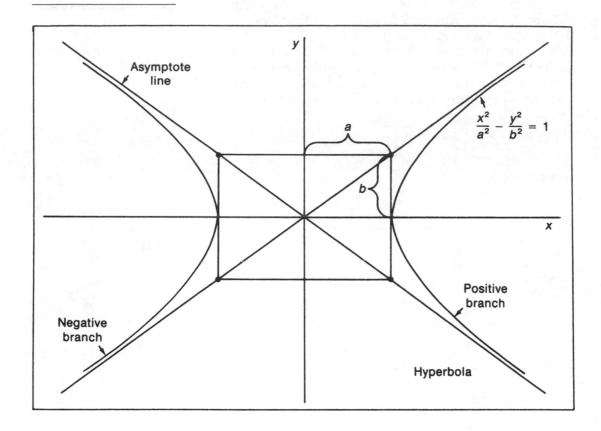

the curve will come closer and closer to the asymptotes, but it will never actually touch them. (See Figure 13-8.)

★*Conic Sections*

5.   Relation to cones

These four curves are called conic sections because they can be formed by the intersection of a plane with a right circular cone. (See Figure 13-9.) If the plane is perpendicular to the axis of the cone, the intersection will be a circle. If the plane is slightly tilted, the result will be an ellipse. If the plane is parallel to one element of the cone, the result will be a parabola. If the plane intersects both nappes of the cone, the result will be a hyperbola. (Note that a hyperbola has two branches.)

Figure 13-9

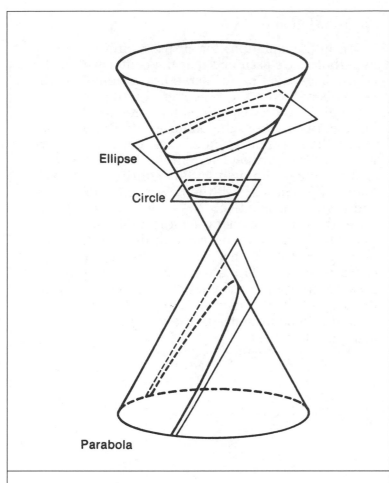

Ellipse

Circle

Parabola

6. General Definition of Conic Sections

It is possible to define ellipses, parabolas, and hyperbolas by one equation. A conic section can be defined as the set of points in a plane such that the distance to a fixed point divided by the distance to a fixed line is a constant. The fixed point is called the focus, the fixed line is called the directrix, and the constant ratio is called the eccentricity of the conic section (abbreviated e). When $e = 1$, this definition exactly matches the definition of a parabola. If e is less than 1, then the conic section is an ellipse. If e is greater than 1, then the conic section is a hyperbola. If the focus is at the point $(0, p)$, the directrix is the line $x = 0$, and the eccentricity is e, then the equation of the conic section can be written as

$$\frac{(x - h^2)}{A} + \frac{y^2}{B} = 1$$

where $h = p/(1 - e^2)$

$A = e^2 p^2/(1 - e^2)^2$

$B = e^2 p^2/(1 - e^2)$

7. Translation of Axes

In the equations for ellipses, circles, and hyperbolas, we assumed that the center was at the origin. The equation for parabolas assumes that the vertex is at the origin. However, it will often be convenient to find equations for conic sections located anywhere in the plane. To do that we use the method of translation of axes. Let's suppose we form a new coordinate system by shifting the x axis h units to the right and by shifting the y axis k units up. We will call the new x axis the x' axis and the new y axis the y' axis. (See Figure 13-10.) (Note the difference between a translation and a rotation. With a rotation we kept the origin at the same place but we changed the direction of the coordinate axes. With a translation we keep the axes pointing in the same direction but we move the origin.) We can convert coordinates in the old system into coordinates in the new system by using the formulas

$$x' = x - h$$
$$y' = y - k$$

Figure 13-10

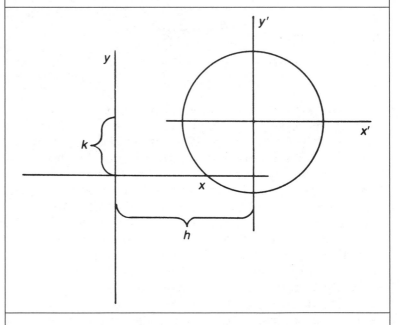

We can convert coordinates in the new system into coordinates in the old system by using the formulas

$$x = x' + h$$
$$y = y' + k$$

Now, suppose we would like to find the equation for a circle with center at the point

(15, 12) and radius 5. Then we perform a coordinate translation like

$$x' = x - 15$$

$$y' = y - 12$$

In the new system the equation of the circle will be very simple:

$$x'^2 + y'^2 = 5^2$$

Now that we know the equation of the circle in the new system, we can use the coordinate translation formulas to calculate the equation of the circle in the old system:

$$(x - 15)^2 + (y - 12)^2 = 5^2$$

In general, the equation of a circle with center at the point (h, k) can be written as

$$(x - h)^2 + (y - k)^2 = r^2$$

The equation of an ellipse with center at the point (h, k) can be written as

$$\frac{(x - h)^2}{a^2} + \frac{(y - k)^2}{b^2} = 1$$

The equation of a parabola with vertex at the point (h, k), focus at the point $(h, k + a)$, and directrix at the line $y = k - a$ can be written as

$$y - k = \frac{(x - h)^2}{4a}$$

The equation of a hyperbola with center at the point (h, k) can be written as

$$\frac{(x - h)^2}{a^2} - \frac{(y - k)^2}{b^2} = 1$$

"Now, you give me a quadratic equation with two unknowns and I will draw a graph of the solution for you," the professor said confidently. "If the graph of the solution is not immediately obvious from the equation, then the trick is to use the right translation of axes to convert the equation into a simple form."

Recordis wrote down the hardest equation he could think of:

$$0.047x^2 - 0.02114xy + 0.0551y^2 - 0.15344x$$
$$- 1.17562y + 6.0625 = 0$$

"No problem," the professor said. However, she gulped when she looked closely at the equation. "I can't solve that equation! It has that $0.02114xy$ term. I don't know how to solve an equation with an xy term in it!"

★*The Pesky xy Term*

"But you said you could solve any equation provided the degree of each term was 2 or less!" Recordis said. "The way I look at it, the term $0.02114xy$ consists of an x to the first power multiplied by a y to the first power, so the term $0.02114xy$ seems to have degree 2."

The professor wrung her hands in deep embarrassment. She had been so confident when she had boasted about her ability to solve quadratic equations that she felt determined to find a way to solve this equation. However, try as she might, she could not find a way. "The solution might or might not be a conic section," she said.

"It's too bad this isn't a trigonometry problem," Trigonometeris said. "If it was, then I'm sure that the trigonometric functions would help you solve it."

The professor struggled with this problem for hours. She had trouble finding even one solution, let alone finding the graph of the complete set of solutions.

That evening Builder came by with a perplexing problem. "I am trying to design a special scoreboard lighting display for the upcoming big game. The scoreboard is made up of lots of little lights. To make different patterns appear, I must decide which lights to light up. That is, I need to know the equations of various curves, such as circles and parabolas. Once I know the equation of a curve, I can calculate the coordinates of the light bulbs I want to light up. You have given me equations for circles, ellipses, parabolas, and hyperbolas. However, all these equations describe figures oriented either vertically or horizontally. You have not told me the equation of a figure tilted with respect to the coordinate axes. For example, I would like to display a parabola tilted at an angle, like this." (See Figure 13-11.)

★The Perplexing Parabola with the Tilted Axis

Figure 13-11

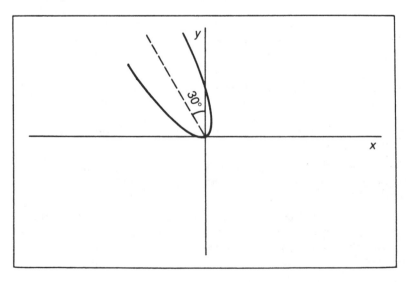

"It makes me very disoriented to look at a crooked parabola like that," Recordis said.

"We can use coordinate rotation," Trigonometeris said. "Let's set up a rotated coordinate system like this." (See Figure 13-12.) "We will rotate the axes by 30°. Then, the equation of the parabola in the rotated coordinate system will be very simple: it will be $y' = x'^2$."

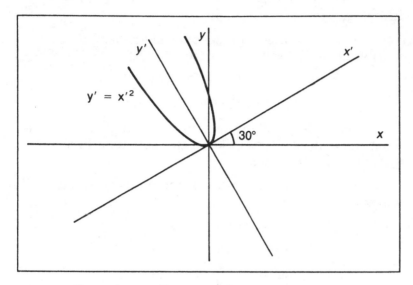

Figure 13-12

"Aha!" the professor said. "Once we know the equation of the parabola in the new system, we can use the rotation formulas

$$x' = x\cos\theta_0 + y\sin\theta_0$$

$$y' = y\cos\theta_0 - x\sin\theta_0$$

to calculate the equation of the parabola in the old system:

$$y' = x'^2$$

$$(y\cos\theta_0 - x\sin\theta_0) = (x\cos\theta_0 + y\sin\theta_0)^2$$

$$y\cos\theta_0 - x\sin\theta_0 = x^2\cos^2\theta_0 + 2xy\cos\theta_0\sin\theta_0$$

$$+ y^2\sin^2\theta_0$$

"In this case we know that the angle of rotation is $\theta_0 = 30° = \pi/6$ rad," the king said. " Since $\cos 30° = \sqrt{3}/2$ and $\sin 30° = \frac{1}{2}$, we can fill in these results. The equation of a parabola with its vertex at origin and the axis tilted by 30° is

$$\frac{3}{4}x^2 + \frac{\sqrt{3}}{2}xy + \frac{1}{4}y^2 + \frac{1}{2}x - \frac{\sqrt{3}}{2}y = 0$$

"Bummer!" Recordis complained. "That equation contains one of those pesky xy terms."

"That doesn't hurt us now because we already know the solution to this equation," Trigonometeris said. "We know the graph of the solution is a parabola tilted by 30°."

The professor was suddenly struck with an idea. "I bet the solution to a two-unknown quadratic equation containing an xy term will indeed be a conic section—but the only difference will be that the conic section's orientation will be rotated away from normal!"

"An interesting guess," the king said. "Let's investigate to see if it is right."

We set up a general second-degree equation involving an xy term.

$$Ax^2 + Bxy + Cy^2 + Dx + Ey + F = 0$$

(A, B, C, D, E, and F are known.)

"Now, let's rotate the axes by an angle θ_0 and see what the equation looks like in the new coordinate system," said the king.

Write x and y in terms of the new coordinates x' and y':

$$x = x'\cos\theta_0 - y'\sin\theta_0$$

$$y = y'\cos\theta_0 + x'\sin\theta_0$$

Find expressions for xy, x^2, and y^2:

$$xy = x'y'\cos^2\theta_0 + x'^2\sin\theta_0\cos\theta_0 - y'^2\sin\theta_0\cos\theta_0 - y'x'\sin^2\theta_0$$

$$x^2 = x'^2\cos^2\theta_0 - 2x'y'\cos\theta_0\sin\theta_0 + y'^2\sin^2\theta_0$$

$$y^2 = y'^2\cos^2\theta_0 + 2x'y'\cos\theta_0\sin\theta_0 + x'^2\sin^2\theta_0$$

Insert these expressions into the original equation:

$$A(x'^2\cos^2\theta_0 - 2x'y'\cos\theta_0\sin\theta_0 + y'^2\sin^2\theta_0)$$
$$+ B(x'y'\cos^2\theta_0 + x'^2\sin\theta_0\cos\theta_0 - y'^2\sin\theta_0\cos\theta_0$$
$$- y'x'\sin^2\theta_0)$$
$$+ C(y'^2\cos^2\theta_0 + 2x'y'\cos\theta_0\sin\theta_0 + x'^2\sin^2\theta_0)$$
$$+ D(x'\cos\theta_0 - y'\sin\theta_0)$$
$$+ E(y'\cos\theta_0 + x'\sin\theta_0) + F = 0$$

After combining all these terms, the equation became

$$x'^2(A\cos^2\theta_0 + C\sin^2\theta_0 + B\sin\theta_0\cos\theta_0)$$
$$+ y'^2(A\sin^2\theta_0 + C\cos^2\theta_0 - B\sin\theta_0\cos\theta_0)$$
$$+ x'(D\cos\theta_0 + E\sin\theta_0) + y'(-D\sin\theta_0 + E\cos\theta_0)$$
$$+ x'y'(-2A\cos\theta_0\sin\theta_0 + 2C\cos\theta_0\sin\theta_0$$
$$+ B\cos^2\theta_0 - B\sin^2\theta_0) + F = 0$$

Recordis's wrist was exhausted after writing that equation. "I think things would be wonderful if we could get rid of that $x'y'$ term."

"We can get rid of the term if we can make this complicated expression go to zero," the professor said.

$$-2A\cos\theta_0\sin\theta_0 + 2C\cos\theta_0\sin\theta_0$$
$$+ B\cos^2\theta_0 - B\sin^2\theta_0 = 0$$

Trigonometeris recognized the formulas for $\cos 2\theta$ and $\sin 2\theta$:

$$(C - A)\sin 2\theta_0 + B\cos 2\theta_0 = 0$$

Now solve for θ_0:

$$\frac{\sin 2\theta_0}{\cos 2\theta_0} = \frac{B}{A - C}$$

$$\tan 2\theta_0 = \frac{B}{A - C}$$

$$2\theta_0 = \arctan\frac{B}{A - C}$$

$$\theta_0 = \frac{1}{2}\arctan\frac{B}{A - C}$$

"Perfect!" the professor said. "The entire $x'y'$ term will vanish if we rotate the axes by an angle

$$\theta_0 = \frac{1}{2}\arctan\frac{B}{A - C}$$

Once we have gotten rid of that bothersome $x'y'$ term, we can use my methods to solve the equation." The professor outlined her new complete method to solve second-degree two-unknown equations.

To solve this equation,

$$Ax^2 + Bxy + Cy^2 + Dx + Ey + F = 0 \qquad (1)$$

1. Calculate the angle of rotation:

$$\theta_0 = \frac{1}{2}\arctan\frac{B}{A - C}$$

(Note that, if $B = 0$, there is no xy term and you will not need to rotate the axes at all.)

2. Let x', y' represent the new axes in the rotated coordinate system. In the new system the equation will be of the form

$$A'x'^2 + C'y'^2 + D'x' + E'y' + F' = 0 \qquad (2)$$

(We could write that equation as

$$A'x'^2 + B'x'y' + C'y'^2 + D'x' + E'y' + F' = 0$$

but we know that $B' = 0$ if we have chosen the angle of rotation correctly.)

You may calculate the new coefficients A', C', D', E', and F' directly by substituting these expressions in the original equation:

$$x' = x\cos\theta_0 + y\sin\theta_0$$
$$y' = y\cos\theta_0 - x\sin\theta_0$$

Or you may use the formulas

$$A' = A\cos^2\theta_0 + C\sin^2\theta_0 + B\sin\theta_0\cos\theta_0$$
$$C' = A\sin^2\theta_0 + C\cos^2\theta_0 - B\sin\theta_0\cos\theta_0$$
$$D' = D\cos\theta_0 + E\sin\theta_0$$
$$E' = -D\sin\theta_0 + E\cos\theta_0$$
$$F' = F$$

If either A' or C' is zero, then the graph of this equation will be a parabola. Suppose that $C' = 0$ (in other words, there is no term containing y'^2). Then, create a new system of coordinates x'', y'' by the translation

$$x'' = x' + \frac{D'}{2A'}$$

$$y'' = y' + \frac{4A'F' - D'^2}{4A'E'}$$

The equation will become

$$A'x''^2 + E'y'' = 0 \tag{3}$$

which can be graphed as a parabola.

If neither A' or C' is zero in Equation (2), then perform the translation

$$x'' = x' + \frac{D'}{2A'}$$

$$y'' = y' + \frac{E'}{2C'}$$

Then the equation can be written as

$$A'x''^2 + C'y''^2 + F'' = 0 \tag{4}$$

If $A' = C'$, then this is the equation of a circle. If A' and C' have the same sign (in other words, they are both positive or both negative), then the equation will be the equation of an ellipse. If A' and C' have opposite signs, then the equation will be the equation of a hyperbola.

★The Graph of the Tilted Ellipse

"It is a very complicated method," the king said, "but we should be able to execute the method if we follow it carefully, one step at a time." The professor was very proud of the result.

Now we had to solve Recordis's equation:

$$0.0474x^2 - 0.02114xy + 0.0551y^2 - 0.15344x - 1.17562y + 6.0625 = 0$$

First, we needed to identify the coefficients with the letters we had used in the standard formula: $A = 0.0474$; $B = -0.02114$; $C = 0.0551$; $D = -0.15344$; $E = -1.17562$; and $F = 6.0625$. Then we calculated the angle of rotation:

$$\theta_0 = \frac{1}{2}\arctan\frac{B}{A-C}$$
$$= 35°$$

Then we formed the new equation in the rotated coordinate system:

$$0.04x'^2 + 0.062499y'^2 - 0.8x' - 0.875y' + 6.0625 = 0$$

In this equation, we identified $A' = 0.04 = \frac{1}{25}$, $C' = 0.062499 = \frac{1}{16}$, $D' = -0.8 = -\frac{20}{25}$, $E' = -0.875 = -\frac{14}{16}$, and $F' = 6.0625 = \frac{97}{16}$.

Then we used the translation formulas

$$x'' = x' + \frac{D'}{2A'}$$
$$= x' - 10$$

$$y'' = y' + \frac{E'}{2C'}$$
$$= y' - 7$$

and the new equation became

$$\frac{x''^2}{25} + \frac{y''^2}{16} = 1$$

"That's an ellipse, with a semimajor axis equal to 5 and a semiminor axis equal to 4!" the professor said. "That will be easy to draw." (See Figure 13-13.)

"There must be a way to tell in advance what the graph of this type of equation will look like!" Recordis said. He thought a bit and came up with an idea. "When we worked with one-variable quadratic equations, such as $ax^2 + bx + c = 0$, we found that the quantity $b^2 - 4ac$ gave us a clue about the nature of the solutions. When we have the equation

$$Ax^2 + Bxy + Cy^2 + Dx + Ey + F = 0$$

let's calculate the quantity $B^2 - 4AC$. We will call $B^2 - 4AC$ the *discriminant* of the quadratic equation with two variables, just as we call $b^2 - 4ac$ the discriminant of the quadratic equation with one variable. I bet this quantity will give us a clue to the nature of the solution."

"That is total nonsense!" the professor exclaimed. "There is no connection between the

★**The Discriminant**

Figure 13-13

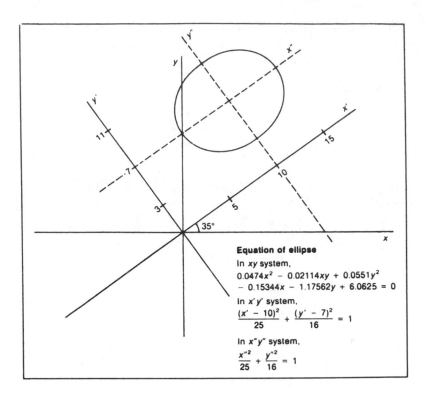

Equation of ellipse

In *xy* system,

$$0.0474x^2 - 0.02114xy + 0.0551y^2 - 0.15344x - 1.17562y + 6.0625 = 0$$

In *x'y'* system,

$$\frac{(x' - 10)^2}{25} + \frac{(y' - 7)^2}{16} = 1$$

In *x"y"* system,

$$\frac{x''^2}{25} + \frac{y''^2}{16} = 1$$

quantity $b^2 - 4ac$ for a one-variable equation and the quantity $B^2 - 4AC$ for a two-variable equation! It is pure coincidence those quantities look the same."

However, to the professor's complete astonishment, Recordis turned out to be right this time. First, we found that the quantity $B^2 - 4AC$ does not change when you rotate the axes by any amount. (For a proof of this rather remarkable fact, see Exercise 18.) In other words, the quantity $B^2 - 4AC$ in Equation (1) will equal the quantity $B'^2 - 4A'C'$ in Equation (2). [However, note that $B' = 0$ because there is no $x'y'$ term in Equation (2).

Recordis explained his plan. "For this equation,

$$A'x'^2 + C'y'^2 + D'x' + E'y' + F' = 0$$

we will define the discriminant as

$$\text{Discriminant} = -4A'C'$$

If A' or C' is zero, then the discriminant = 0 and the curve is a parabola; if A' and C' have the same sign, then the discriminant is negative and the curve is an ellipse; if A' and C' have opposite signs, then the discriminant is positive and the curve is a hyperbola. Therefore, we can make this rule."

> *Consider the Equation*
>
> $$Ax^2 + Bxy + Cy^2 + Dx + Ey + F = 0$$
>
> Calculate the quantity $B^2 - 4AC$. If $B^2 - 4AC$ is 0, then the graph of this equation will be a parabola; if $B^2 - 4AC$ is negative, then the graph will be a circle or an ellipse; and if $B^2 - 4AC$ is positive, then the graph is a hyperbola.

"But this result must be blind luck!" the professor still insisted. She could not figure out how Recordis had become so lucky to be able to guess the rule that would determine in advance the nature of the solutions to the complicated quadratic equation with two unknowns. She was sure that in general she was the best person in the kingdom when it came to discovering new scientific ideas, but she had to concede that Recordis was the best person when it came to discovering innovations designed to save work.

Our study of coordinate rotation gave Builder the idea for a spinning ride. The children sat on a spinning disk. The amount the disk had rotated at a given time was θ, which was found from the formula:

$$\theta = \omega t$$

where t was time and ω was the angular frequency. (Builder could adjust the value of ω; a larger value would make the ride spin faster.) Builder drew a coordinate grid on the ride. If a child sat at the point (x, y) on this grid, the coordinates in the frame of reference of the ride would not change. However, according to the (x', y') coordinate system used by those of us outside the ride, the position of the child at time t is found from the equations

$$x' = x\cos(\omega t) + y\sin(\omega t)$$

$$y' = y\cos(\omega t) - x\sin(\omega t)$$

Although the children liked the ride at first, some of the more adventuresome ones began to think it was too tame. Builder decided to add some excitement by creating a spherical ride that would spin vertically as well as horizontally. "I represent points in space with an xyz coordinate system," he explained to us. "First, I rotate the x and y axes about the z axis by an angle of θ. Then, I rotate the x and z axes about the y axis by an angle of ϕ."

"It makes me dizzy just thinking about it," Recordis shuddered.

The professor worked out the formula for coordinate rotation when there were two separate

Three-Dimensional Rotations and Virtual Reality

rotations, about both the z axis and the y axis. x', y', and z' are the coordinates in the system used by those of us outside the ride, and x, y, and z are the coordinates relative to the ride.

$$x' = x\cos\theta\cos\phi + y\sin\theta\cos\phi - z\sin\phi$$

$$y' = y\cos\theta - x\sin\theta$$

$$z' = z\cos\phi + x\cos\theta\sin\phi + y\sin\theta\sin\phi$$

The children enjoyed this ride (as long as Builder was careful to make sure that it did not spin too fast). However, Builder was already thinking of a new kind of ride. "I'll call it a virtual reality ride," he explained. "My computer will keep track of the coordinates of a group of objects in their own frame of reference. Then the game player will enter this virtual world, and we'll be able to adjust the display so that the objects appear as they are seen by the player. If the virtual player turns to the side, the computer will rotate the coordinates of all the objects so that they appear as they would if the player really had turned to the side. And if the player looks up or down, the computer can rotate the coordinates of all the objects up or down."

One again we were astounded at the unexpected directions in which our study of trigonometry had taken us.

Note to CHAPTER 13

- You have probably played computer games that allow you to change your viewpoint as described here. Perhaps one day you will be able to write your own games that do this. For an example of a computer program that creates a simple object and rotates it in three dimensions, see this book's web page at *http://www.spu.edu/~ddowning/easytrig.html*.

Exercises

In Exercises 1 to 14, you are given the coordinates of some points in the initial xy system. Calculate the new coordinates (x', y') in a coordinate system formed by rotating the axes by an angle θ_0:

	x	y	θ_0
1.	1	1	45°
2.	1	1	90°
3.	1	1	135°
4.	1	1	180°
5.	1	1	225°
6.	10	15	12°
7.	12	0	30°
8.	12	0	80°
9.	12	0	120°

10.	0	25	25°
11.	0	25	330°
12.	26	68	78°
13.	10	6	63°
14.	91	28	57°

15. In the text we derived the formulas that tell how to convert coordinates in the old xy system to the new $x'y'$ system. Derive the formulas that tell how to convert the (x', y') coordinates to the (x, y) coordinates.

★16. Use a translation of axes to convert this equation:

$$Ax^2 + Cy^2 + Dx + Ey + F = 0$$

into this form:

$$\frac{(x-h)^2}{G} + \frac{(y-k)^2}{L} = 1 \quad \text{or} \quad \frac{x'^2}{G} + \frac{y'^2}{L} = 1$$

★17. Use a translation of axes to convert this equation:

$$Cy^2 + Dx + Ey + F = 0$$

into the equation of a parabola with vertex at the origin.

★18. Show that the quantity $B^2 - 4AC$ in the equation

$$Ax^2 + Bxy + Cy^2 + Dx + Ey + F = 0$$

is the same as the quantity $B'^2 - 4A'C'$ when you rotate the axes of the coordinate system.

★19. Derive the polar coordinate equation for a general conic section. Put the focus at the origin and put the directrix at the line $r\cos\theta = -a$.

Draw the graphs of the solutions to the equations in Exercises 20 to 25.

20. $17.0528x^2 + 23.9472y^2 + 5.7851xy - 400 = 0$

21. $19.7186x^2 + 21.28142y^2 + 8.863xy - 400 = 0$

22. $95.25x^2 + 85.75y^2 - 16.454xy - 8100 = 0$

23. $54.75x^2 - 35.75y^2 - 156.75xy - 8100 = 0$

24. $8.849x^2 + 4.1508y^2 - 1.710xy - 36 = 0$

25. $-8.608x^2 + 3.608y^2 + 4.446xy - 36 = 0$

26. Use a coordinate rotation to graph the equation $xy = 1$.

27. Can you think of circumstances in which the graph of the solutions to the equation $Ax^2 + Bxy + Cy^2 + Dx + Ey + F = 0$ will not be a conic section?

Calculate the angle of rotation you would use if you were to draw the graphs of the equations in Exercises 28 to 33. Without actually drawing the graph, determine the nature of the graph.

28. $16x^2 + 2xy - 18y^2 + 12x - 14y - 56 = 0$

29. $33x^2 + 53xy + 62y^2 - 61x - 80y - 81 = 0$

30. $x^2 + 10xy + y^2 - 5x - 10y - 44 = 0$

31. $4x^2 - xy + 8y^2 - 2x - y - 45 = 0$

32. $x^2 + 6xy + 9y^2 + x + 2y - 81 = 0$

33. $4x^2 + 16xy + 16y^2 + 3x + 6y - 144 = 0$

□★34. If you need to graph the equation

$$Ax^2 + Bxy + Cy^2 + Dx + Ey + F = 0$$

it would be very difficult to perform all the calculations by hand. Write a program that reads in the values for the coefficients, prints a message describing the nature of the graph of the solution, and then draws the graph.

★35. Find a simple formula for $x'^2 + y'^2 + z'^2$ for the case of three-dimensional rotation about two axes (see page 210):

$$x' = x\cos\theta\cos\phi + y\sin\theta\cos\phi - z\sin\phi$$

$$y' = y\cos\theta - x\sin\theta$$

$$z' = z\cos\phi + x\cos\theta\sin\phi + y\sin\theta\sin\phi$$

14

Spherical Trigonometry

The Royal Astronomer, who also had the responsibility of being Royal Navigator, asked Recordis to draw a map to help him plan a new journey. "The Southsea islanders have put lighthouses on three main islands: North, South, and West Islands," he explained. "They have carefully measured the distance between the islands, and they have carefully noted the shortest course between each pair of islands. South Island and West Island are located exactly along the equator, and North Island is directly north of South Island."

"It will be a piece of cake to draw the map," Recordis said. Even though he always complained about work, he still appreciated the fact that people turned to him for important graphing jobs because they knew he was the best in the kingdom. "You have described a right triangle."

★*Mysterious Triangles and Riddles*

The astronomer told Recordis the distances between the three pairs of islands and the sizes of the other two angles in the right triangle. Recordis started drawing the map. (See Figure 14-1.)

Figure 14-1

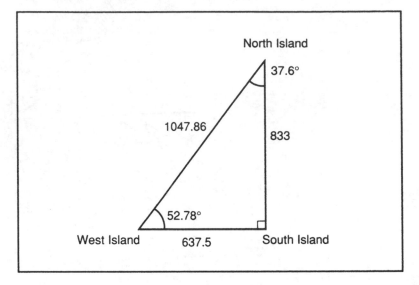

However, when he started to double-check the numbers, something puzzling happened. He applied the Pythagorean theorem, but he found there was a slight discrepancy:

$$\sqrt{833^2 + 637.5^2} = 1048.95$$

But the astronomer has supplied the figure 1047.86 for the distance from West Island to North Island.

"Are you sure that these measurements are accurate?" Recordis asked.

"Positive," the astronomer said. "I have carefully double-checked, and in each case the measurements provided by the islanders are extremely accurate."

Recordis was still puzzling over this problem when he received an even bigger shock. He added together the three angles provided by the astronomer:

$$37.6 + 52.78 + 90 = 180.38°.$$

"No! This cannot be!" Recordis screamed. "We know that the three angles in a triangle must always add up to 180°! That is true for any triangle!"

The astronomer insisted that the measurements were accurate. Fortunately, the professor arrived at that moment and she agreed to investigate the mystery. Meanwhile, the astronomer described another one of his long sea journeys to Recordis.

"I left Startingpoint Island, on the equator, and sailed 2000 kilometers directly west. Then I turned at an angle of 72.8° and sailed for 5304.6 kilometers, then

I turned directly south and sailed for 5000 kilometers until I returned to Startingpoint Island. During each leg of the journey, my guide painstakingly made sure that we were always following the shortest course between the two points." (See Figure 14-2.)

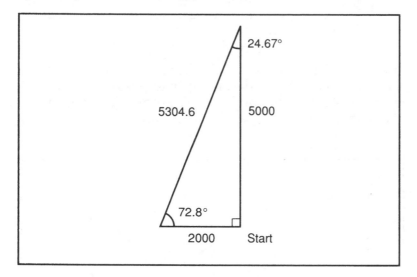

Figure 14-2

"This makes another right triangle, which is nice because right triangles are easier to deal with than other types of triangles," Recordis said. However, his relief quickly turned to despair when the same two problems surfaced, only this time they were much worse: the Pythagorean theorem did not work, and the sum of the three angles was 187.5°, again greater than 180°.

"What is going on?" Recordis cried. "We have never had these problems before!" Again, the astronomer insisted that the measurements were extremely accurate.

"The only clue we have is that the problem becomes much worse with a much bigger triangle," the professor said.

"But if the Pythagorean theorem is wrong, and the three-angles-sum-to-180° rule is wrong, then we have to throw out everything we have done!" Recordis cried in anguish.

"That would be very distressing," the professor agreed.

We all suffered a sleepless, disturbed night. To make matters worse, the next morning Pal received a little riddle book from an anonymous stranger. Normally he liked riddles, but now he was crying because the first riddle was very difficult.

"Let me try," said Recordis. He read the riddle:

"A hunter left his home and walked two miles directly south. Then he walked three miles directly east. At that point he saw a bear. He then turned directly north and walked two miles, until he reached his home. What color was the bear?"

"There is no information about the color of the bear at all!" Recordis complained.

For lack of anything better to do, the professor tried to make a diagram of the hunter's course. "Aha!" she said. "The course described is impossible. You would not return to your starting point if you followed the path described." (See Figure 14-3.)

(See Figure 14-3.)

Figure 14-3

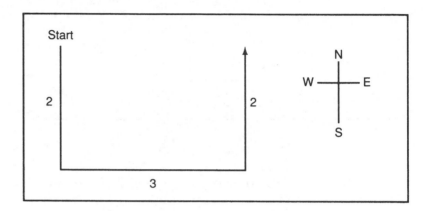

"Did you like my little riddle?" a passing stranger asked. He turned to us, and we recognized the gremlin!

"You gave that riddle book to Pal!" the king accused him.

"There is an answer to the riddle," he laughed, "and the course described is possible." Then he slipped away, but his scornful laughter rang in our ears.

We all realized that our triangle problems from yesterday, and this strange riddle, put into jeopardy all of the work we had done with triangles. Trigonometeris, naturally, took matters worse than anyone else.

"I thought we fully understood triangles by now," he said.

Pal was crying so hard that the king went to fetch his beachball to cheer him up. However, before he tossed the ball to the giant, the king suddenly stopped and stared at the ball.

"I think we are missing something," he said slowly. "The Earth is round, like a ball. We have looked at triangles drawn on flat pieces of paper, but I suspect that a triangle drawn on a sphere will behave differently."

It took a moment for the significance of what he had said to sink in. Then Trigonometeris became very excited. "A new type of trigonometry! We will call it *spherical trigonometry*, since it is about triangles on spheres."

"We should use the term *plane trigonometry* for the work we have done up to now," the professor said.

"Don't imply that trigonometry is plain and ordinary!" Trigonometeris exclaimed.

"That's not the type of plane I meant," the professor said. "A plane is a flat surface, like a table top. All of the triangles we have looked at before yesterday were plane triangles."

"Does this help us answer the riddle?" the king asked.

"It still isn't possible to follow the path described in the riddle and end up at the starting point," Recordis said, "unless you do something strange like start at the North Pole."

"That's it!" the king said. "If you start at the North Pole, travel south, then east, and then turn due north, you will come back to your starting point!" (See Figure 14-4.)

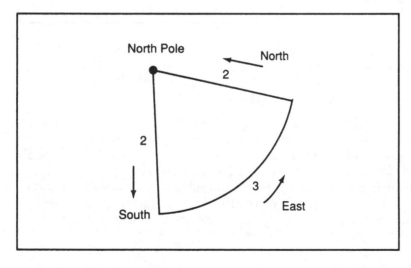

Figure 14-4

"White bear!" Pal shouted happily, realizing that if the hunter started at the North Pole, he must have seen a white polar bear.

We set to work investigating the properties of spherical triangles. "First, we need a system to keep track of locations on a sphere, similar to the rectangular (*xy*) coordinate system we use to identify locations on a plane," the professor said.

"Since a sphere exists in three dimensions, we will need a three-dimensional coordinate system,"

Recordis said. "We used this type of system before when we investigated equations with three variables in algebra. We added the *z* axis to measure distances above or below the origin. With the *xyz* system, we can identify any point in three-dimensional space." (See Figure 14-5.)

Figure 14-5

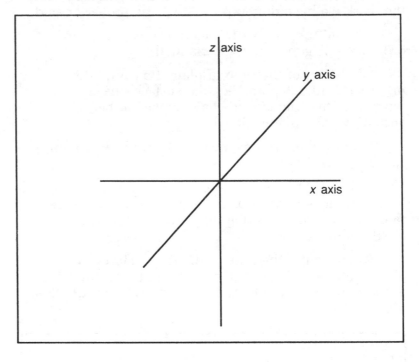

"At the moment we don't need to identify all possible points in three-dimensional space," the professor said. "We only need to identify those points that are located along the sphere."

The astronomer explained, "Actually, I find that I can keep track of my position by using just two coordinates. As long as I stay on the equator, I just need to keep track of how far I have travelled east from the starting point. I call this distance the *longitude*. If I then turn and sail north or south of the equator, then I need to keep track of the distance north or south. I call this the *latitude*." (See Figure 14-6.)

Figure 14-6

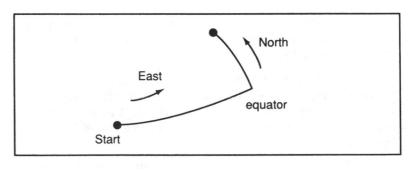

"What is the equator?" Trigonometeris asked.

"The equator is the great circle that goes all the way around the Earth, halfway between the North and South Poles," the astronomer explained.

"What's so great about it?"

The astronomer tried to explain what a great circle was, but Trigonometeris only became more confused. Fortunately, at that moment the Royal Baker happened to come by carrying a large ball of butter, which had been made unusually large for an unusually special occasion.

The astronomer took out a large knife. "Watch what happens when I cut the ball of butter." As the astonished baker watched, the astronomer cleanly sliced part of the ball off. Then he took the slice and put it on the table, asking Recordis to trace out the shape formed by the cross section of the slice.

"It makes a circle!" Recordis said in astonishment. (See Figure 14-7.)

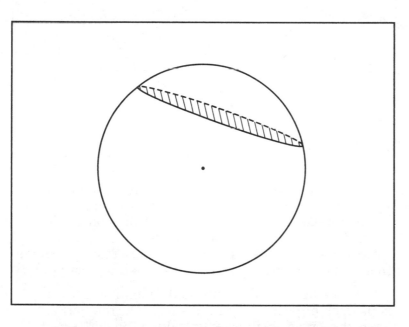

Figure 14-7

"Whenever you cut a sphere with a straight cut, as I have just done, then a circle is formed. However, this type of circle is called a small circle, because the diameter of the circle is less than the diameter of the sphere. Watch this. . . ."

The astronomer replaced the slice, and the ball of butter again formed a perfect sphere. Then he sliced the ball again, only this time he was careful to make sure that his knife passed through the center of the ball. The ball split into two equal hemispheres. Again Recordis traced the pattern formed, and again he found that it was a circle. This time the diameter of the circle was the same as the diameter of the sphere.

The professor suddenly realized what was happening. "Whenever a plane crosses a sphere so that the plane passes through the center of the sphere, then the intersection of the plane and the sphere forms a great circle. A great circle is bigger than any other circle that you form by crossing the sphere with a plane that does not cross the center of the sphere." (See Figure 14-8.)

Figure 14-8

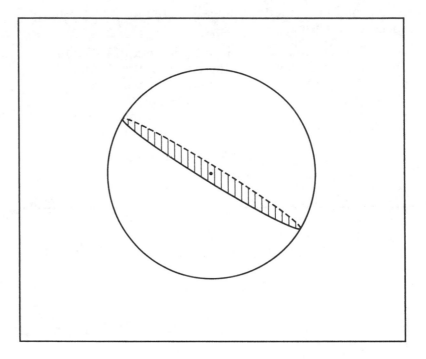

We all took turns slicing the ball of butter, and we were able to convince ourselves that every time our knife crossed the center of the ball, the great circle formed had the same diameter as the sphere itself. However, every time our knife did not cross the center then the circle formed was smaller than a great circle.

The astronomer continued to explain how he kept track of positions on the surface of the Earth. "I have chosen an arbitrary point along the equator to represent the starting point for measuring longitude."

"How do you measure the distances?" the king asked.

★Latitude and Longitude

"The distance can be measured in kilometers, but I find it convenient to measure distance by giving them in relation to circles. I have followed your suggestion where a complete circle measures 360° or 2π radians. Therefore, 180° means to sail halfway around the world, 90° means to sail one quarter of the way around the world, and so on." He showed us an example of how to reach the point with longitude 20° and latitude 15°. (See Figure 14-9.)

Figure 14-9

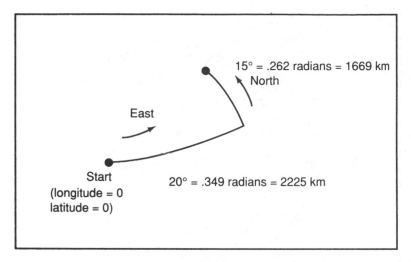

15° = .262 radians = 1669 km

North

East

Start
(longitude = 0
latitude = 0)

20° = .349 radians = 2225 km

"We need some symbols for latitude and longitude," Recordis said. "It would be too confusing to use L, because they both start with L."

"We will have to use Greek letters again," the professor said. She suggested θ (theta) for longitude, and φ (phi) for latitude.

The king issued a proclamation:

Any point on the surface of the Earth can be identified by two coordinates. First, choose a point along the equator to be the starting point (latitude = 0, longitude = 0).

* The longitude of point X tells how far you need to travel eastward along the equator from the starting point to reach point E, which is the point on the equator directly north or south of the point X. If θ is the longitude in radians, then the distance of travel is θr, where r is the radius of the Earth.

* The latitude of point X tells how far you need to travel from point E to point X (this trip will be directly north or directly south). If φ is the latitude, then the distance of travel is φr. (See Figure 14-10.)

It is customary to express latitude and longitude in degrees, although they can be measured in radians if it is more convenient in a particular circumstance.

Figure 14-10

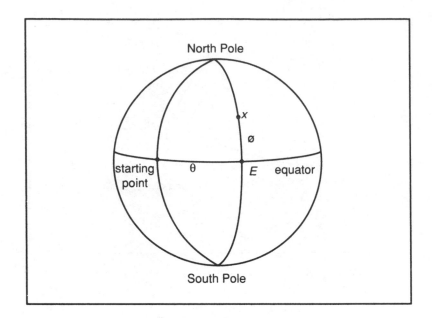

We used the term *circle of longitude* to refer to a great circle consisting of all points with the same longitude (and, as you can see, the points on the opposite side of the Earth whose longitude is 180° greater). Half of such a great circle, consisting of points with the same longitude, is called a *meridian*. We used the term *circle of latitude* (or *parallel* of latitude) for a circle consisting of all points with the same latitude. Figure 14-11 shows a globe with several circles of latitude and circles of longitude.

Figure 14-11

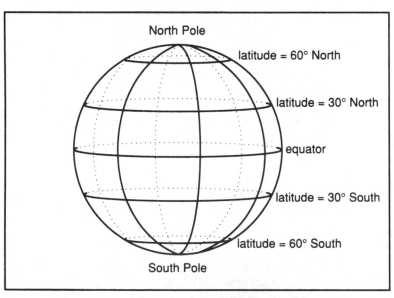

Note that a circle of latitude (other than the equator) is not a great circle. A circle of latitude is formed by cutting the Earth's sphere with a plane parallel to the plane of the equator. Such a plane will not pass through the center.

"I would feel much more comfortable if we had a way to convert latitude and longitude coordinates into *xyz* coordinates," Recordis said. In order to humor him, the professor agreed to help find the *xyz* coordinates for a point on the Earth's surface with latitude φ and longitude θ.

"We should put the origin at the center of the Earth," the professor said. "We will have the *x* axis pointing to the starting point (latitude = 0, longitude = 0), the *y* axis pointing toward latitude = 0, longitude = 90°, and the *z* axis pointing to the North Pole." (See Figure 14-12.)

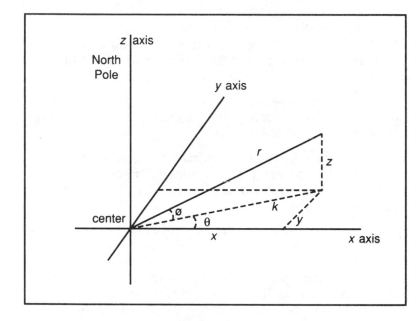

Figure 14-12

"We know that the distance from the origin to any point on the surface is *r*, where *r* is the radius of the Earth," Recordis said. "Everyone knows that the Earth is a perfect sphere."

(The astronomer did not want to interrupt them, but he did whisper to the rest of us, "Actually, the Earth is not exactly a perfect sphere, but it is so close to being a sphere that for practical purposes we can treat it as if it is one.")

From the diagram we could see:

$$z/r = \sin\phi$$

$$k/r = \cos\phi$$

$$y/k = \sin\theta$$

$$x/k = \cos\theta$$

Putting these together, we obtained three formulas.

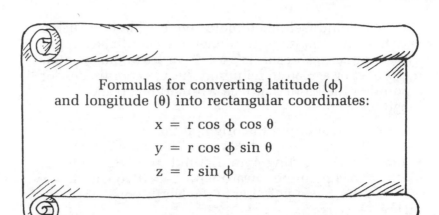

Formulas for converting latitude (ϕ)
and longitude (θ) into rectangular coordinates:

$$x = r \cos \phi \cos \theta$$

$$y = r \cos \phi \sin \theta$$

$$z = r \sin \phi$$

Note that these formulas are similar to the polar coordinate transformation formulas in Chapter 11. If you want to turn x, y, and z coordinates into latitude and longitude, see Exercise 18.

"Now we need to return to our main business at the moment: investigating the properties of spherical triangles," the professor said.

★Spherical Triangles

"We should make spherical triangles as much as possible like ordinary plane triangles," Recordis said. "Therefore, a spherical triangle will have three vertices and three straight sides."

"I am afraid it is more complicated than that," the king realized. "The sides of a spherical triangle cannot be straight lines. It is impossible to travel in a straight line between two points on the surface of the Earth without tunnelling or swimming under the surface. Therefore, the sides of a spherical triangle must be curved arcs."

"Then the side of a spherical triangle will be the arc that connects the two vertices," Recordis said.

"Which arc?" the astronomer asked. "There are many different arcs that connect two points on a sphere." The astronomer needed to convince Recordis of the fact, so he sent a message to the baker asking for another ball of butter. He marked two points on the ball, and then cut a slice out of the ball. He was careful that the knife cut both of the marked points. He did this several times, and each time the cut indicated a different arc connecting the two marked points.

"We should choose the shortest arc," the king decided. After considerable investigation, we found that the shortest route between two points was the route that was part of a great circle.

The professor decided that we should investigate the spherical triangle representing the astronomer's course, which had sides of 5000, 2000, and 5304.6

kilometers. (See Figure 14-2.) She started to sketch the triangle on the butter ball. She calculated that the radius of the ball was 1.6×10^{-7} times the radius of the Earth, so she told Recordis to divide 5000, 2000, and 5304.6 by 1.6×10^{-7} so she would know how long to make the sides of the triangle on the butter ball.

"We need a way to measure the length of the sides of a spherical triangle that works for spheres of all sizes, so we don't need to change all the numbers like that," Recordis complained.

"I have an idea," the king said. "We will measure the side by calculating the ratio of the length of the side to the radius of the circle." The astronomer told us that the radius of the Earth was 6375 kilometers, so we calculated:

Length of side (in kilometers)	Length of side (in radians)
5000.0	5000.0/6375 = 0.7843
2000.0	2000.0/6375 = 0.3137
5304.6	5304.6/6375 = 0.8321

"This is just like radian measure for angles," Trigonometeris realized. (See Figure 14-13.)

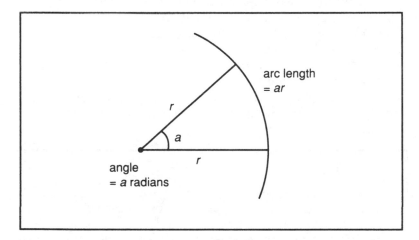

Figure 14-13

The professor carefully drew the graph on the butter ball. However, it had become warmer in the room, and by the time she finished the ball was beginning to melt. "We will have to draw spherical triangles on paper," Recordis said. "It is difficult to draw a perspective diagram to create the illusion of three-dimensional space on a two-dimensional piece of paper, but we have no other choice."

We drew a graph of a spherical triangle, using a, b, and c to represent the lengths of the sides, measured in radian measure. "That means that the lengths of the sides are *ar*, *br*, and *cr*, expressed in kilometers, if *r* is the radius of the sphere in kilometers," Recordis reminded us.

"With plane triangles, we found it convenient to use capital letters to represent the angles, with *A* being the angle opposite side *a*, and so on," Trigonometeris reminded us. (See Figure 14-14.)

Figure 14-14

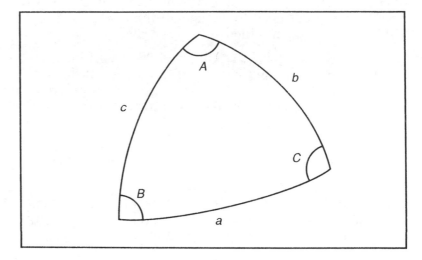

"How do we measure the angles between two curves?" Recordis asked.

The professor had an idea. "Remember that each curve is a part of a great circle, which is formed by a plane cutting the sphere and passing through the center. Therefore, the angle between two sides is the angle between the two planes." Builder quickly constructed a device with a hinge to illustrate the angle between two glass planes with circles painted on them. (See Figure 14-15. Such an angle is called a *dihedral angle*.)

Figure 14-15

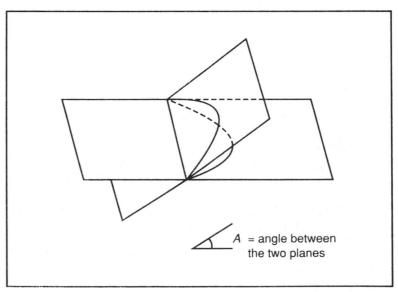

We realized that spherical triangles presented a danger of confusion, because both the sides and angles are measured in angular measure. (This problem does not occur with plane triangles, since the lengths of the sides are measured in linear measures such as inches or kilometers.) Recordis convinced us that it would be a little less confusing if we measured the angles of spherical triangles in degrees, but then measured the sides in radians. That way we could simply multiply the radian measure of a side by r to obtain the length of the arc for that side. We will follow this rule unless there is a particularly compelling reason why it is more convenient to measure the sides in degrees.

"Now we need some formulas that connect the sides and the angles," the professor said.

"I suggest we make the problem simpler by first looking at spherical triangles that contain a right angle," Recordis suggested. "We found that it was easier to investigate plane right triangles before we looked at other types of plane triangles."

We drew a sketch of a spherical right triangle, imagining that our sketch represented a big spherical triangle on the surface of the Earth. Since the relations between the sides and angles are the same no matter where the triangle is located, Recordis insisted that we figure out the most convenient location for the triangle before we drew it. After a long discussion, we decided that one side of the triangle (side b) should be along the equator, and that side a should be along a circle of longitude, which goes directly north from the equator. That meant that the right angle was between sides a and b, and that side c was the side opposite the right angle. (See Figure 14-16.) Note that we drew the great circle containing side b (the equator) and the great circle containing side c. The dihedral angle between the two planes containing these circles was angle A of the triangle. We decided to put vertex A at the starting point —that is, at the point on the equator with longitude 0.

★ Spherical Right Triangles

Figure 14-16

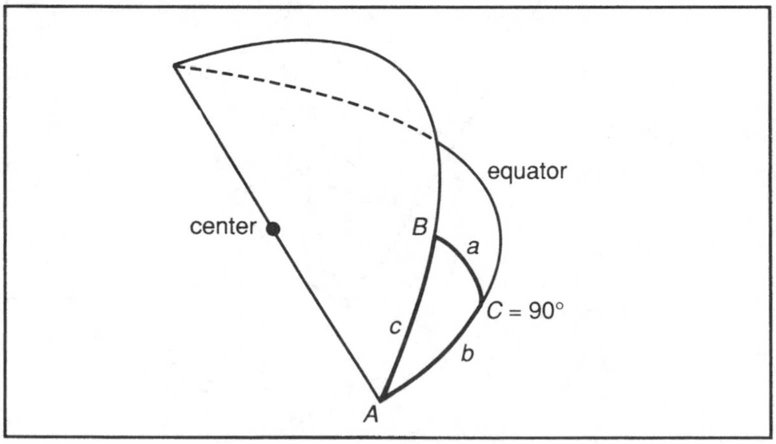

We spent a long time trying to figure out what to do next. Finally, Trigonometeris said, "As surprising as it may seem, I think Recordis actually came up with the correct idea a while back. We should find the rectangular coordinates for the three vertices."

Because of the choice we had made for the location of the triangle we could see that vertex A was located at the point where latitude = 0 and longitude = 0. Vertex C had latitude 0, since it was on the equator. b measured the distance (in radians) that you must travel along the equator to reach vertex C, so we realized that b was the longitude of vertex C. Since vertex B was directly north of vertex C, it must have the same longitude. Since a measured the distance (in radians) by which vertex B was north of the equator, that meant that a was the latitude of vertex B.

Now we could use the formulas we had derived earlier (page 224) to find the three rectangular coordinates of vertex B (with longitude b and latitude a):

$$x = r\cos a\cos b$$

$$y = r\cos a\sin b$$

$$z = r\sin a$$

(See Figure 14-17. Recall that we put the origin at the center of the Earth, with the x axis pointing to the starting point and the z axis pointing to the North Pole. The arrows show the direction of the x and y axes. The arrows were not actually drawn at the origin because that would have made the diagram too cluttered.)

"We still can't make much progress unless we can somehow find some plane right triangles," the professor said. Just as she was saying this, she realized

Figure 14-17

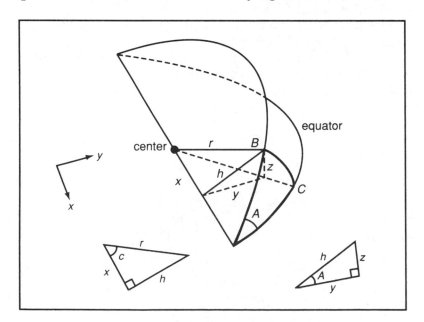

that we could form some plane right triangles if we drew more line segments on the diagram. First, we drew line segments connecting the origin to the three vertices of the triangle. Each of these line segments had length r. Then we drew a vertical line segment from vertex B down to the plane containing the equator; this line segment represented the z coordinate of vertex B. Then we drew two more line segments (labeled h and y in Figure 14-17).

"Let's write the Pythagorean theorem for the plane right triangle with legs y and z and hypotenuse h," Trigonometeris said. (Note that Figure 14-17 contains a sketch of two plane right triangles separated from the main part of the diagram for clarity.)

$$h^2 = y^2 + z^2$$

"We know expressions for y and z, but what about h?"

"We need to stare at the diagram to see if we can find any other right triangles containing h," the professor said.

"I see one!" Trigonometeris said. "It has legs of length x and h, and a hypotenuse of r. The angle opposite side h measures c radians."

Therefore: $h = r \sin c$

We substituted the expressions for y, z, and h into the equation $h^2 = y^2 + z^2$:

$$r^2 \sin^2 c = r^2 \cos^2 a \sin^2 b + r^2 \sin^2 a$$

We cancelled the r^2's and then used some trigonometric identities to simplify the equation:

$$1 - \cos^2 c = \cos^2 a (1 - \cos^2 b) + \sin^2 a$$
$$1 - \cos^2 c = \cos^2 a - \cos^2 a \cos^2 b + \sin^2 a$$
$$1 - \cos^2 c = 1 - \cos^2 a \cos^2 b$$
$$\cos^2 c = \cos^2 a \cos^2 b$$
$$\cos c = \cos a \cos b$$

"This is an amazingly simple formula!" Trigonometeris said. "It relates the three sides of a spherical right triangle, just as the Pythagorean theorem relates the three sides of a plane right triangle. We just have to remember that we use radian measure to measure the lengths of the sides, and that c is the side opposite the right angle."

We tested out the formula on the astronomer's triangle with sides of 0.8321, 0.7843, and 0.3137 radians:

$$\cos 0.8321 = \cos 0.7843 \cos 0.3137$$
$$0.6733 = 0.7079 \times 0.9512$$

"It works!" exclaimed the professor.

Trigonometeris then suggested that we should look for a formula for $\sin A$. "If we have a plane right triangle, then we know that $\sin A = a/c$, if c is the hypotenuse and a is the side opposite angle A."

"Matters will be more complicated with spherical triangles," Recordis warned. However, we could quickly see from one of the plane right triangles in Figure 14-17 that $\sin A = z/h$. Since we already had expressions for z and h we could substitute:

$$\sin A = \frac{r \sin a}{r \sin c}$$

$$\sin A = \frac{\sin a}{\sin c}$$

"Now let's find an expression for $\cos A$," Trigonometeris continued. Again using the triangle in Figure 14-17, we could see that $\cos A = y/h = (r \cos a \sin b)/(r \sin c)$. We decided to use the equation $\cos c = \cos a \cos b$ to substitute. Then:

$$\cos A = \frac{\cos c \sin b}{\cos b \sin c}$$

which we realized we could write like this:

$$\cos A = \frac{\tan b}{\tan c}$$

"Again, this is analogous to a plane right triangle, where $\cos A = b/c$," Trigonometeris suggested.

"We have lots of momentum now," the professor said excitedly. We could easily find a formula for $\tan A$:

$$\tan A = \frac{\sin A}{\cos A} = \frac{\dfrac{\sin a}{\sin c}}{\dfrac{\tan b}{\tan c}}$$

$$= \frac{\sin a \tan c}{\sin c \tan b}$$

$$= \frac{\sin a \sin c \cos b}{\sin c \cos c \sin b}$$

$$= \frac{\sin a \cos b}{(\cos a \cos b) \sin b}$$

(using the equation $\cos c = \cos a \cos b$)

$$= \frac{\sin a}{\cos a \sin b}$$

$$\tan A = \frac{\tan a}{\sin b}$$

The professor realized that we could use exactly the same procedures to come up with corresponding formulas for $\sin B$, $\cos B$, and $\tan B$. We made a summary of our results:

Spherical Right Triangles

Three-Sides Formula:

$$\cos c = \cos a \cos b$$

Side/Two Adjacent Angles Formula:

$$\cos c = \text{ctn } A \text{ ctn } B$$

Formulas for angle A	Corresponding formulas for angle B
$\sin A = \dfrac{\sin a}{\sin c}$	$\sin B = \dfrac{\sin b}{\sin c}$
$\cos A = \dfrac{\tan b}{\tan c}$	$\cos B = \dfrac{\tan a}{\tan c}$
$\tan A = \dfrac{\tan a}{\sin b}$	$\tan B = \dfrac{\tan b}{\sin a}$
$\sin A = \dfrac{\cos B}{\cos b}$	$\sin B = \dfrac{\cos A}{\cos a}$

"In these formulas it is imperative to recall that the capital letters represent the angles in the triangle, and the lower case letters represent the three sides of the triangle, measured in radians," the professor reminded us. "Also remember that the right angle is opposite side c."

We called the first formula the three-sides formula since it relates the lengths of the three sides. The second formula is derived in Exercise 38. We derived the last two formulas when the professor asked, "I wonder if $\cos B$ is equal to $\sin A$, since that is the way it works for plane right triangles." Starting from the equation $\cos B = \tan a / \tan c$, she found:

$$\cos B = \frac{\tan a}{\tan c}$$

$$= \frac{\sin a \cos c}{\cos a \sin c}$$

$$= \frac{\sin a \cos a \cos b}{\sin c \cos a}$$

$$= \sin A \cos b$$

Therefore:

$$\sin A = \cos B / \cos b$$

We verified that these formulas worked for the triangles that the astronomer had given us (See Exercise 16).

"Now we can finally put these equations to use," the astronomer said. "Suppose that I am at point 1 (latitude = ϕ_1, longitude θ_1) and I wish to sail to point 2 (latitude ϕ_2, longitude θ_2). I wish to travel along the shortest course possible, which, as we have seen, is the great circle path. However, it is very difficult to determine specifically what that course is, since I usually do not have a convenient spherical butter ball available."

"The problem is easy if the two points are on the equator," the professor said helpfully. "Since the equator is a great circle, then the shortest path between the two simply involves sailing along the equator."

"I know that," the astronomer said. "It is also easy to sail between two points of the same longitude, since you merely sail directly north (or south) until you reach the point you wish. Recall that circles of longitude are all great circles."

"It also should be easy to sail between two points of the same latitude," Recordis suggested. "You would then sail directly east or west until you reach the destination."

"But that would not be the shortest course!" the astronomer said. "Recall that circles of latitude are *not* great circles (with the exception of the equator itself). Therefore, if you follow a course of constant latitude, you will not be sailing along the shortest possible course between the two points."

"I see that this is a tricky problem," the professor said. "Perhaps it would help if we could draw a spherical triangle with these two points as vertices."

"Where will we put the other vertex?" Recordis asked. "We need to choose some place very convenient." As it turned out, the North Pole was the most convenient place to put the third vertex of the triangle. (See Figure 14-18.)

★*Sailing Along the Shortest Course*

Figure 14-18

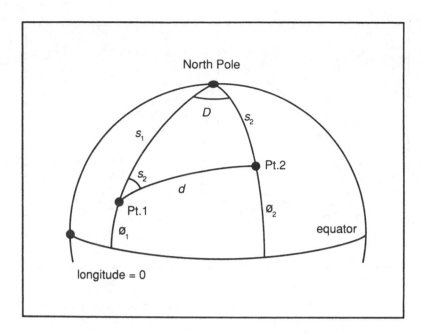

"By 'convenient' I mean a point that is convenient theoretically. I am not suggesting that we actually travel to the North Pole," the professor said.

We used d to represent the unknown distance between the two points along the great circle course, and s_1 and s_2, the complements of the latitude of the two points, made up the other two sides of the spherical triangle: $s_1 = 90° - \phi_1$; $s_2 = 90° - \phi_2$. The angle D was the difference in longitude between the two points: $D = \theta_2 - \theta_1$.

"Since circles of longitude are great circles, and the optimal path between points 1 and 2 is a great circle, we know that we have drawn a spherical triangle."

"It is not a right triangle, though, so we cannot use our right triangle formulas. We need a more general formula, like the law of sines and the law of cosines for plane triangles," Trigonometeris said.

We looked back at Recordis's notes to see how we had found the law of sines and the law of cosines. We decided to try an analogous procedure for the case of spherical triangles. We started with an arbitrary spherical triangle (sides a, b, and c; angles A, B, and C). Then we drew an arc from vertex B that was perpendicular to side b; this formed two right spherical triangles. (See Figure 14-19.) We let h represent the length of this arc. Then, using the formulas from page 231, we wrote:

$$\sin A = \frac{\sin h}{\sin c} \qquad \sin C = \frac{\sin h}{\sin a}$$

★**Law of Sines and Law of Cosines**

Figure 14-19

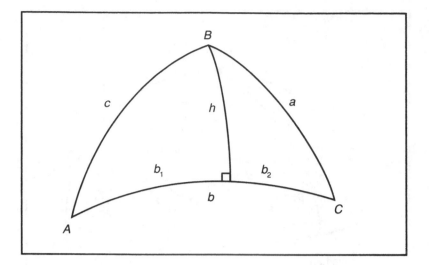

(For the right triangle on the left, c is the side opposite the right angle; for the right triangle on the right, a is the side opposite the right angle.) We solved both of these equations for $\sin h$ and then set them equal:

$$\sin A \sin c = \sin C \sin a$$

"This is almost exactly like the law of sines for plane triangles!" Trigonometeris suggested. "Let's write it down."

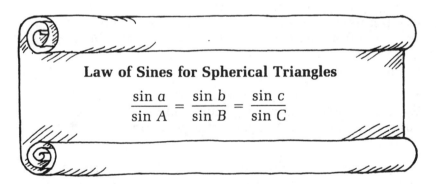

Law of Sines for Spherical Triangles

$$\frac{\sin a}{\sin A} = \frac{\sin b}{\sin B} = \frac{\sin c}{\sin C}$$

(Note that we can use a similar procedure to show that we can include $\sin b/\sin B$ in the formula.)

"Now we should look for the law of cosines for spherical triangles," Recordis suggested.

From the three sides equation for spherical right triangles, we knew these equations were true for the two right spherical triangles in Figure 14-19:

$$\cos c = \cos h \cos b_1 \qquad \cos a = \cos h \cos b_2$$

Solving the second equation for $\cos h$ and then substituting into the first equation:

$$\cos c = \frac{\cos a \cos b_1}{\cos b_2}$$

Since $b_1 = b - b_2$, we know:

$$\cos(b_1) = \cos(b - b_2) = \cos b \cos b_2 + \sin b \sin b_2$$

Substituting:

$$\cos c = \frac{\cos a (\cos b \cos b_2 + \sin b \sin b_2)}{\cos b_2}$$

$$\cos c = \cos a \cos b + \cos a \sin b \tan b_2$$

Using another equation from page 222, we could substitute $\tan b_2 = \cos C \tan a = \cos C \sin a / \cos a$:

$$\cos c = \cos a \cos b + \cos a \sin b \cos C \left(\frac{\sin a}{\cos a} \right)$$

$$\cos c = \cos a \cos b + \sin a \sin b \cos C$$

"How elegant!" Trigonometeris said. "We will call this the law of cosines for sides of a spherical triangle."

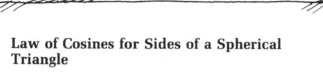

Law of Cosines for Sides of a Spherical Triangle

$$\cos c = \cos a \cos b + \sin a \sin b \cos C$$

"It even works for spherical right triangles," the professor said. "Suppose C is a right angle. Then $\cos C = 0$, and we have:

$$\cos c = \cos a \cos b$$

which is the same as the three sides equation for right triangles."

"Now we can finally solve the problem," the astronomer said. "Suppose that we are at point 1 (latitude 20°, longitude 205°), and we wish to sail to point 2 (latitude 32°, longitude 239°). Then we set up our spherical triangle." (See Figure 14-20.)

Figure 14-20

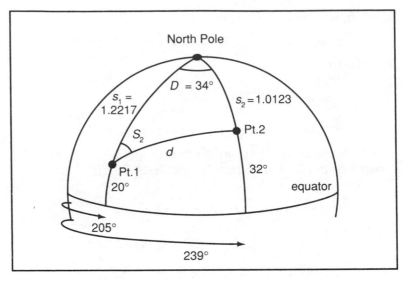

"We can see that side s_1 is 90°—latitude of point 1, if we measure it in degrees," the professor said. "Likewise, side s_2 is 90°—latitude of point 2."

"Angle D is simply the difference between the longitudes of the two points," the king contributed. Then we had:

$$s_1 = 90 - 20 = 70° = 1.2217 \text{ radians}$$

$$s_2 = 90 - 32 = 58° = 1.0123 \text{ radians}$$

$$D = 239 - 205 = 34°$$

We used the law of cosines for spherical triangles:

$$\cos d = \cos s_1 \cos s_2 + \sin s_1 \sin s_2 \cos D$$
$$\cos d = \cos(1.2217)\cos(1.0123)$$
$$\qquad + \sin(1.2217)\sin(1.0123)\cos(34°)$$
$$\cos d = 0.3420 \times 0.5299 + 0.9397 \times 0.8481 \times 0.8290$$
$$\cos d = 0.8419$$
$$d = \arccos 0.8419 = 0.5700 \text{ radians}$$

From this we could calculate the shortest (great circle) distance between the two points:

$$\text{distance} = dr = 0.5700 \times 6375 = 3634 \text{ kilometers}$$

Next, we used the law of sines for spherical triangles to find angle S_2:

$$\sin S_2 = \frac{\sin s_2 \sin D}{\sin d} = \frac{\sin(1.0123)\sin(34°)}{\sin(0.5700)}$$
$$= \frac{0.8481 \times 0.5592}{0.5396}$$
$$= 0.8789$$
$$S_2 = \arcsin 0.8789 = 61.51°$$

"Just what I need to know," the astronomer said. "When I leave point 1, I will set my course at an angle 61.5° east of north." (The professor had misgivings, because she realized that sin 118.49° was also equal to 0.8789, but in this case it was clear from the diagram that the correct angle must be 61.51°. This ambiguity inherent in the sine function caused us more grief later.)

"I have just realized one very depressing fact," Recordis said. "All of the maps I have ever drawn are wrong! I have drawn maps on flat pieces of paper, containing plane triangles, but since the world is really a sphere, I should have been drawing spherical triangles.

We were very worried about this fact. However, after examining several maps of cities it became clear that there wasn't a practical problem. "If you have a spherical triangle whose sides are very small, relative to the entire sphere, then the spherical triangle is almost exactly the same as a plane triangle," the king said. "We do not need to worry about the spherical Earth when we draw maps of small areas. However, I can see that if we try to draw a map of a very large area of the Earth, then we will not be able to ignore the fact that the Earth is a sphere."

Early the next morning we received a letter from Mrs. O'Reilly of the Carmorra Beachfront Hotel. The hotel was so successful that she was planning to expand to a new location, and she asked us for some advice. "When planning a beach hotel, it is very important to know about the angle of altitude of the sun," she wrote. "Could you tell me how to calculate the angle of altitude of the sun at noon on the first days of winter, spring, summer, and fall?"

★The Altitude of the Sun at the Beach

The astronomer told us that this was an easy problem. "Let's calculate z, the angle between the sun and the zenith (the point directly above the observer's head). Once we know the zenith angle, we can easily calculate the altitude:

$$\text{altitude} = 90° - z$$

"On both the first day of fall and the first day of spring, the sun is directly overhead at the equator at noon. If the hotel is at latitude ϕ, then the zenith angle at noon equals ϕ. Because the Earth's axis is tilted by i = 23.45°, on the first day of summer the sun is overhead at points 23.45° north of the equator, so its zenith angle at noon is $\phi - i$. On the first day of winter, the sun is overhead at points 23.45° south of the equator, so its zenith angle at noon is $\phi + i$." (See Figure 14-21. Note that the sun is so far away that the light rays from the sun that hit two different parts of the Earth can be treated as if they are parallel.)

Figure 14-21

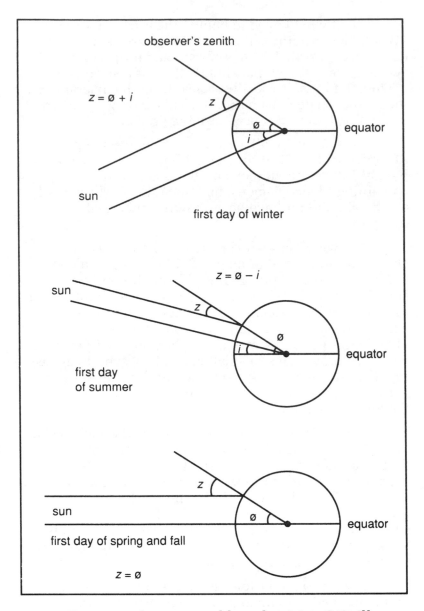

However, the next problem that Mrs. O'Reilly asked us to solve was not easy. She asked us to give her a formula that would give the altitude of the sun at various times of the day. This one had us stumped.

"If only this was a problem involving spherical triangles!" Recordis moaned.

We worked on this problem all day and all night. We watched the sun rise, move across the sky, and then set. Builder built us a device (called a *sextant*) that we could use to measure the altitude angle of the sun at a particular time, but we did not know how to write a formula that would allow us to calculate the altitude. We watched the stars become visible, and we saw stars rise and set throughout the night. "All of the stars move, except for that one," Recordis noticed long after midnight. He pointed to a bright star.

"That is the North Star," the astronomer said. "It would be directly overhead if you were at the North Pole. All of the other stars move in circles around it."

"I see," Recordis said. "All of the stars are glued to the inside of a giant ball that turns around the Earth. The North Star is the pivot around which the ball turns."

"That is not true at all!" the astronomer blustered.

"Even so, I think it will help to imagine that the stars are glued to the inside of a giant ball," the professor said, beginning to get an idea. "We will call this ball the *celestial sphere*. We can see the north celestial pole, the point that is directly above the Earth's North Pole. I suggest that we use the term *celestial equator* for the great circle across the celestial sphere that is above the Earth's equator." (See Figure 14-22.)

★*The Celestial Sphere*

Figure 14-22

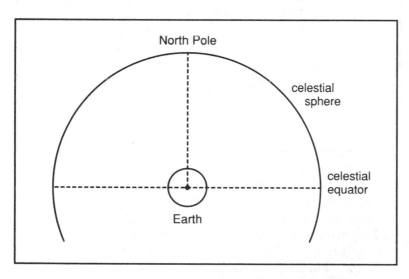

"We can measure positions along the celestial sphere, just as we measure positions along the surface of the Earth with latitude and longitude," the king realized.

"We better not use the terms latitude and longitude, or we will become confused with positions on the Earth," Recordis warned.

The astronomer, who had more contact with the rest of the world than the other members of the royal court, suggested two terms that he had heard: *declination* and *right ascension*. Declination is used to measure the angular distance between a star and the celestial equator. Stars on the equator have a declination of 0, and the North Star has a declination of approximately 90°. Declination is the celestial equivalent of latitude. Right ascension is the celestial equivalent of longitude. The right ascension of a star tells you how far you need to travel along the celestial equator to reach the point directly south (or north) of that star. Just as with longitude, we needed to decide on an arbitrary starting point for right ascension. The astronomer suggested the point that is the location of the sun on the first day of spring. For example, to find a star at right ascension 24° and declination 8°, you need to travel 24° from the starting point along the celestial equator, and then travel 8° directly north. Of course, you cannot really travel along the celestial sphere, but you could use these instructions to aim a telescope at the star. (See Figure 14-23.)

Figure 14-23

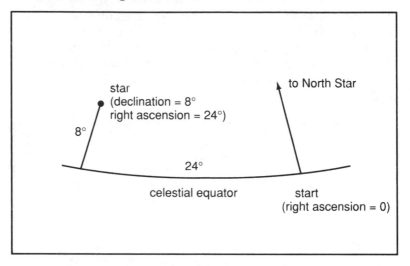

The astronomer quickly promised to measure the positions of many bright stars and make a table showing their right ascensions and declinations. He even said that he could measure the right ascension and declination of the sun, since we realized that the sun also moved along the celestial sphere even if it was so bright that none of the other stars were visible during the day. Recordis then thought that the problem was solved, but the astronomer sadly explained that Mrs. O'Reilly was asking a different question.

"If we measure the declination of a star, then we can subtract to find the angular distance between that star and the North Star. That distance will not change through the night. However, the angle between the star and the horizon will change. That is what Mrs. O'Reilly wants to know."

A while later Recordis became tired. He lay down on his back on the patio, with his head pointing north. He watched as a star passed the zenith directly overhead and then continued moving to the west.

Recordis continued to watch the North Star, the zenith, and the other star. He briefly fell asleep, and then suddenly woke with a start.

"I have been working with spherical triangles too much," he complained." I was having a dream where the North Star, the zenith, and the other star formed a spherical triangle." (See Figure 14-24. Remember that this figure shows what you would see if you lie on your back with your head pointing directly north.)

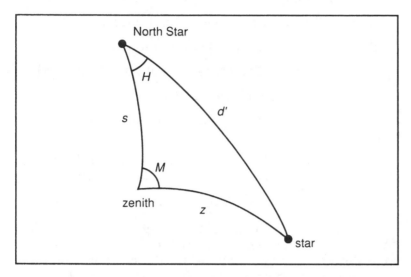

Figure 14-24

"Let's label the sides of this triangle to see what they represent," Trigonometeris suggested. "Side z represents the angular distance between the star and the zenith. If we can calculate z, then we can find the altitude of the object: altitude $= 90° - z$."

"Side d' represents the angular distance between the North Star and the star," the professor said. "If we know the declination (d) of the star, then $d' = 90° - d$."

"Side s represents the angular distance between the North Star and the zenith," the astronomer said. "Suppose ϕ is your latitude. Then $s = 90° - \phi$."

"I have been watching this triangle throughout the night," the king said. "The zenith and the North Star stay in the same place, but the star itself is always moving. Side d' always has the same length, but the angle H is always changing." (Note that we do not always follow the tradition of naming an angle with the capital form of the letter representing the opposite side, since sometimes we can think of names that better help us remember what the angle means.)

"I can calculate the angle H without too much problem," the astronomer said. "I call it the *hour angle* of the star. It depends on two things—the right ascension of the star and what time it is (i.e., how much the celestial sphere has turned)."

"That is everything we need to know," the professor said. "We have a spherical triangle with two known sides, and we know the angle between those two sides. To find the third side we use the law of cosines:

$$\cos z = \cos s \cos d' + \sin s \sin d' \cos H$$

where

z = zenith angle = 90° − star's altitude

$s = 90° - \phi$, where ϕ = latitude of observer

$d' = 90° - d$, where d = declination of star

H = hour angle of star (see note at end of chapter)

M = direction to look to see star

(We didn't need to use angle M at the moment, but we figured we might need it later.)

Since the cosine of an angle is equal to the sine of its complement, we could rewrite this equation:

$$\sin(\text{altitude}) = \sin\phi \sin d + \cos\phi \cos d \cos H$$

"That solves the problem!" Trigonometeris said. "All the astronomer needs to do is come up with a table showing the declination and the hour angle of the sun at a particular time on a particular date, and then we can use this formula to find the altitude."

We worked an example for the first day of summer (where the sun's declination was 23°) during a time in the late afternoon when the sun's hour angle was 75°. If Mrs. O'Reilly built the hotel at latitude 20°, the sun's altitude would be:

$$\sin(\text{altitude}) = \sin 20° \sin 23°$$
$$+ \cos 20° \cos 23° \cos 75°$$
$$= 0.3575$$
$$\text{altitude} = 20.9°$$

If she built the hotel at latitude 50°:

$$\sin(\text{altitude}) = \sin 50° \sin 23°$$
$$+ \cos 50° \cos 23° \cos 75°$$
$$= 0.4525$$
$$\text{altitude} = 26.9°$$

"This is interesting," the astronomer said. "During the late afternoon during northern hemisphere summer, the sun seems to be at a higher altitude if you are farther north. At noon, of course, the sun would be at a lower altitude if you were farther north." The astronomer promised to do the necessary work to find the altitude of the sun at other times. (See Exercise 33.) "This formula works for all other stars as well," the astronomer said. "This will help me very much in my work."

We were interrupted by a distress radio call. "It's my assistant!" the astronomer cried. "He has just radioed me that he is on a long sea voyage and he has become lost! He does not know his latitude!"

"Can he see any stars?" the professor asked.

The astronomer spoke into the radio, and the assistant reported that he could.

"No problem," the professor said. "As long as you know the altitude and direction of a star of known declination, we can solve this spherical triangle for the unknown part to find the latitude. (See Figure 14-24.) This time we know d', z, and M, and we are trying to solve for s."

"This will be more difficult, though," Trigonometeris said worriedly. "We cannot use the law of cosines, because the angle that we know is not between the two sides that we know."

"We can solve for angle H, using the law of sines," the king said.

$$\frac{\sin d'}{\sin M} = \frac{\sin z}{\sin H}$$

$$\sin H = \frac{\sin z \sin M}{\sin d'}$$

"Now that we know two angles in the triangle, we can solve for the third angle, can't we?" Trigonometeris asked.

"For plane triangles the sum of the angles is 180°, but that is not true for spherical triangles," Recordis reminded him. "That is what started all of this trouble in the first place."

This proved to be another difficult problem. We continued working on it sleeplessly throughout the night. Recordis decided to write down everything that he did know. He wrote a formula and then double-checked it to make sure that it worked:

Law of Cosines

$$\cos C = -\cos A \cos B + \sin A \sin B \cos c$$

"There is no minus sign in the law of cosines," the professor said.

"I needed to put the minus sign to make the formula work," Recordis said. (See Exercise 40.)

The professor looked more closely at the formula and realized that the sleepy record keeper had written it backward. "You forgot that we use capital letters to stand for angles, and lower case letters to represent the sides!"

The king took an interest in the discussion. "It would be very helpful if this formula would work, even if it was discovered by mistake," he said. We checked a few more examples and found that the formula did work. (Later we were able to prove the formula to be true in general. See Exercise 42.) We called this the law of cosines for angles, and the previous version of the formula we called the law of cosines for sides.

Formulas for Spherical Triangles

Law of Cosines for Sides:

$$\cos c = \cos a \cos b + \sin a \sin b \cos C$$

Law of Cosines for Angles:

$$\cos C = -\cos A \cos B + \sin A \sin B \cos c$$

The professor said we should investigate the general case of a spherical triangle where a, A, b, and B were known, and c and C were unknown. (See Figure 14-14.)

"The law of cosines for sides by itself won't help us find $\cos c$, since $\cos C$ is unknown. Likewise, the law of cosines for angles by itself won't help to find $\cos C$, since $\cos c$ is unknown," Recordis reasoned. "Maybe we can put the two together."

$$\cos c = \cos a \cos b + \sin a \sin b$$
$$\times (-\cos A \cos B + \sin A \sin B \cos c)$$
$$\cos c = \cos a \cos b - \sin a \sin b \cos A \cos B$$
$$+ \sin a \sin b \sin A \sin B \cos c$$
$$\cos c (1 - \sin a \sin b \sin A \sin B)$$
$$= \cos a \cos b - \sin a \sin b \cos A \cos B$$
$$\cos c = \frac{\cos a \cos b - \sin a \sin b \cos A \cos B}{1 - \sin a \sin b \sin A \sin B}$$

The astronomer's hapless assistant had radioed us that he had observed the star Arcturus to have an altitude of 52.33° (so the zenith angle was $z = 90 - 52.33 = 37.67°$, and its direction was $M = 97.02°$ west of north.) The astronomer performed some measurements to determine that the declination of Arcturus was 19.33°, giving us $d' = 90 - 19.33 = 70.67°$. (See Figure 14-25.)

Figure 14-25

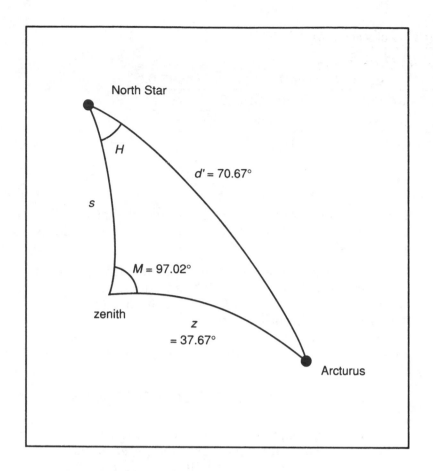

"First, we use the law of sines to calculate angle H," the professor said.

$$\frac{\sin d'}{\sin M} = \frac{\sin z}{\sin H}$$

$$\sin H = \frac{\sin 37.7° \times \sin 97.02°}{\sin 70.67°}$$

$$= 0.6432$$

"This brings us to an annoying little problem," Trigonometeris said. "H could be either 40° or 140°."

"We had the same problem with plane triangles when we knew two sides and one angle other than the one between the two known sides," the professor said. "We called that the ambiguous case, and we found we needed additional information to identify the triangle." (See Chapter 7.)

"There is a big difference between $H = 40°$ and $H = 140°$!" the astronomer said. "I am sure my assistant is bright enough to tell the difference between these two cases without even having to measure H." The assistant radioed back that H was an acute angle. Therefore, it could not be 140°, so it must be 40°. Now we identified the correct parts to use in the formula from page 244:

$$\cos s = \frac{\cos d' \cos z - \sin d' \sin z \cos H \cos M}{1 - \sin d' \sin z \sin H \sin M}$$

$$= \frac{\cos 70.67° \cos 37.67° - \sin 70.67° \times \sin 37.67° \cos 40° \cos 97.02°}{1 - \sin 70.67° \sin 37.67° \sin 40° \sin 97.02°}$$

$$= \frac{(0.3310 \times 0.7915) - [0.9436 \times 0.6111 \times 0.7660 \times (-0.1222)]}{1 - (0.9436 \times 0.6111 \times 0.6428 \times 0.9925)}$$

$$= 0.500$$

$$s = 60°$$

Therefore, we could radio to the assistant that his latitude was 90 − 60 = 30°, and we were finally able to get some sleep.

The next day the professor put together a summary on how to solve for the unknown parts of a spherical triangle. We had to do a little more work to describe all of the possibilities:

★*Solving Spherical Triangles*

Solving Spherical Triangles

1. You must know three parts of a spherical triangle to solve for the other three parts. As is traditional, we use capital letters (A, B, and C) to represent the angles, and lower case letters (a, b, and c) to represent the sides, where side a is opposite angle A, and so on.

2. If you know the three angles, then you may solve for the sides by using the law of cosines for angles:

$$\cos c = \frac{\cos C + \cos A \cos B}{\sin A \sin B}$$

3. If you know the length of two sides (a and b) and the size of the angle between these two sides (C), then you can solve for the third side (c) by using the law of cosines for sides:

$$\cos c = \cos a \cos b + \sin a \sin b \cos C$$

4. If you know the length of one side (c) and the two angles next to that side (A and B), then use the law of cosines for angles:

$$\cos C = -\cos A \cos B + \sin A \sin B \cos c$$

5. If you know the length of the three sides, then use the law of cosines for sides:

$$\cos C = \frac{\cos c - \cos a \cos b}{\sin a \sin b}$$

6. Suppose you know two sides (a and b) and the size of one angle other than the one between those two sides (for example, suppose you know angle A). Then use the law of sines to find angle B:

$$\sin B = \frac{\sin b \sin A}{\sin a}$$

This is the ambiguous case. In general, there will be two possible values of B, one greater than 90° and one less than 90°. You will need additional information to determine which possibility is correct for your situation. Once you know the value of B, find side c from this formula, which comes from combining the law of cosines for sides with the law of cosines for angles:

$$\cos c = \frac{\cos a \cos b - \sin a \sin b \cos A \cos B}{1 - \sin a \sin b \sin A \sin B}$$

Then use the law of sines to find angle C.

7. Suppose you know two angles (A and B) and a side other than the one between those two angles (for example, suppose you know side a). Then use the law of sines to find side b:

$$\sin b = \frac{\sin B \sin a}{\sin A}$$

(However, this is another ambiguous case; in general there are two possible values of b.) Once a, b, A, and B are known, use the formula from rule 6.

There are some other formulas that you may use to solve for the parts of a spherical triangle. (See Exercises 43–46.)

Note that these rules are similar to the rules given for plane triangles in Chapter 7. Plane triangles have no rule corresponding to rule 2, since knowing the three angles of a plane triangle does not give you any information about its size. They also have no rule corresponding to rule 7, since it is easy to calculate the third angle of a plane triangle if you know the other two angles.

* * *

After that, every time we saw the stars we thought with amazement how trigonometry had been the crucial discovery that allowed us to predict their positions.

● It is customary to measure latitude, longitude, and declination in terms of degrees/minutes/seconds, as described on page 11. However, for calculation purposes it is more convenient to use decimal degrees, as we have done in this chapter.

● The starting point for longitude measurements is totally arbitrary. However, you must be consistent: Once you have chosen a starting point, you need to use the same starting point all the time. By custom, the meridian that passes through the Greenwich Observatory at London, England, has zero longitude. There still is one more question: In which direction should the longitude be measured? The formulas given on page 224 work if longitude is measured to the east; so, for example, Paris has longitude 2°, Tokyo 140°, Los Angeles 242°, New York 286°, and Dublin 354°. Mathematically this is most convenient, because then the y axis points to 90° longitude. However, according to custom, only longitudes from 0 to 180° are measured in this direction, with the designation east included: Paris is at 2° east and Tokyo is at 140° east. Longitudes in the other half of the Earth are customarily given as degrees west of Greenwich. For example, Dublin is 6° west, New York is 74° west, and Los Angeles is 118° west. A westward-measured longitude (W) can be converted into an eastward-measured longitude (E) from the formula: $E = 360° - W$.

● This chapter has been written from the viewpoint of a northern hemisphere observer, since most of the people on Earth happen to live in the northern hemisphere. (The exact location of Carmorra itself remains a mystery). The same principles apply to southern hemisphere observers. References to the North Pole should be changed to read "South Pole," and references to the North Star should be changed to read "south celestial pole," which is a location on the celestial sphere although it is not marked by a bright star. The formulas given in the chapter work if latitudes and declinations north of the equator are treated as positive numbers and latitudes and declinations south of the equator are treated as negative numbers.

● The hour angle of an object can be found from this formula: $H = t - RA$, where RA is the right ascension of the object, and t is the local sidereal time. Local sidereal time measures how much the celestial sphere has turned. Suppose that the stars with right ascension zero are currently lined up with the circle that passes from your zenith to the point on your horizon directly south. (Then these stars are said to be on the meridian.) The time when this occurs is called sidereal time zero. About one hour later, the entire celestial sphere has rotated by 15° to the west, and stars with right ascension 15° are now on the meridian. This time is called sidereal time 15°. At this time, the stars at right ascension zero have an hour angle of 15 − 0 = 15°. The celestial sphere continues to rotate throughout the night (and day), with the local sidereal time constantly increasing until it returns to zero when the stars with right ascension zero reappear on the meridian.

For our purposes, it is convenient to measure right ascension and sidereal time in degrees. However, customarily these are measured in units of hours, where 24 hours = one complete circle = 360°, or 1 hour = 1/24 of a complete circle = 15°. This type of hour is slightly different than the hour we commonly use to represent time, since the celestial sphere makes one complete rotation in 23 hours 56 minutes, measured in common (solar) hours.

Figure 14-26 shows the view if you were high above the North Pole looking down. Suppose the celestial equator had numbers painted along it, giving the right ascension in hours. For the observer shown, the figure shows sidereal time 165° = 11 hours. (Note that the sidereal time is the same at all points of the same longitude, but it is different at points with other longitudes.) At this time, for

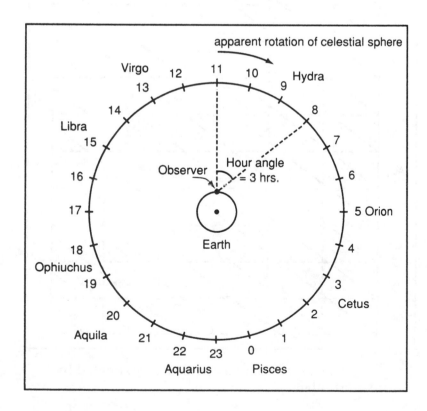

Figure 14-26

example, stars with right ascension 120° = 8 hours have an hour angle of 165 − 120 = 45° = 3 hours, and stars with right ascension 300° = 20 hours have an hour angle of 165 − 300 = −135° = 225° = −9 hours = 15 hours. Note that negative hour angles can be converted to positive numbers by adding 360° or 24 hours. The diagram also indicates the location of a few constellations that are located along the celestial equator.

Of course, the celestial sphere does not really rotate at all; its apparent rotation occurs because of the turning of the Earth.

Figure 14-27 contains a chart allowing you to find the approximate local sidereal time at night by finding the date along the left side of the chart, then finding the local standard time at the top of the chart, and reading the sideral time (in hours) from the diagonal lines.

Figure 14-27

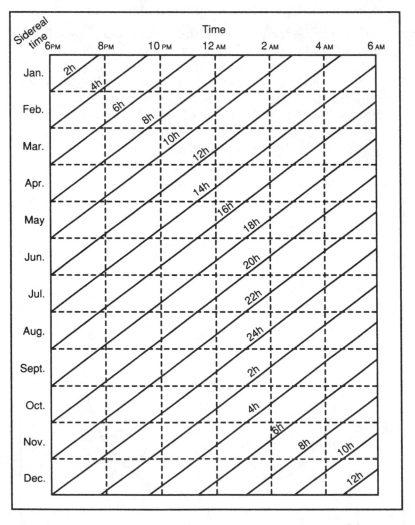

The sidereal time can be approximated by this formula:

$$ST = 12 + \frac{366.25CT}{365.25} + \frac{24D_v}{365.25}$$

where ST is the sidereal time, CT is the local clock time, and D_v is the number of days since the vernal equinox (which takes place about March 21 each year). Both ST and CT are measured on a scale from 0 to 24. If the formula gives a value greater than 24, subtract 24 to make sure the result is between 0 and 24. The value of CT is zero at midnight and 12 at noon. (If daylight savings time is in effect, then subtract one hour. For example, if it is 6 AM daylight time, then CT is 5. You also may need to adjust the formula if your longitude is different from the standard longitude for your time zone.)

• Navigators can use observations of the sun or stars to determine their latitude, as described here. However, you cannot determine longitude by observing stars unless you also have a clock set to a reference time zone that allows you to determine how much the celestial sphere has rotated. Celestial navigation methods in practical use typically involve observing the altitude of two or more stars; see a book on navigation for more details.

• The *azimuth* of a star tells you in which direction to look in order to see it. It is customary to measure azimuth by defining north to be 0, east to be 90°, south 180°, and west 270°. The angle M in the triangle in Figure 14-24 gives the azimuth if the object is in the eastern half of the sky; if the object is in the western half of the sky, as shown in this example, then the azimuth equals $360° - M$.

Exercises

For Exercises 1 to 15, solve for the missing parts of the spherical triangles in Figure 14-28, if possible. In each case the angles are measured in degrees and the sides are measured in radians. (Write a computer program to help if you wish.)

16. Consider the two spherical right triangles provided by the astronomer in Figures 14-1 and 14-2. Verify that these triangles satisfy the ten equations for spherical right triangles.

17. Calculate the distance between the origin and the point with coordinates $x = r\cos\phi\cos\theta$, $y = r\cos\phi\sin\theta$, $z = r\sin\phi$.

18. Suppose you are given the x, y, and z coordinates of a point on the surface of the Earth. Find formulas for the latitude and longitude.

For Exercises 19 to 23, find the rectangular coordinates for the points with given longitude and latitude. ($r = 6375$)

Figure 14-28

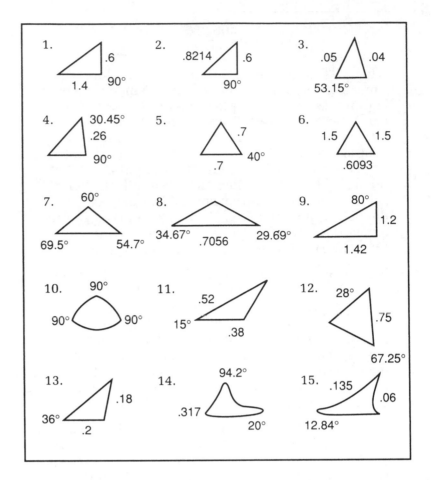

	Latitude	Longitude
19.	0	0
20.	0	90
21.	90	0
22.	30	90
23.	0	270

For Exercises 24 to 28, find the latitude and longitude of the points with the given rectangular coordinates.

24. $x = -1888.3$ $y = -5188$ $z = -3187.5$

25. $x = 553.5$ $y = 3139.1$ $z = -5520.9$

26. $x = -2441.8$ $y = 4229.3$ $z = 4097.8$

27. $x = -4229.3$ $y = 2441.8$ $z = 4097.8$

28. $x = -1090.2$ $y = 1888.3$ $z = 5990.5$

29. Suppose you travel between two points at the same latitude (ϕ) separated by D in longitude. (a) Find a formula for the distance if you travel along the great circle course between the two points. (b) Find a formula for the distance if you travel along a course at a

constant latitude. (c) Consider the course of the hunter shown in Figure 14-4. Is the second leg of his journey a great circle course or a constant latitude course? (d) Suppose you are travelling between Chicago and Rome, both at latitude 42° and separated by 100° of longitude. Calculate the distance along both the great circle course and the constant latitude course.

30. In the spherical triangle in Figure 14-29, suppose D_1, S_2, and s_1 are all known. Describe a procedure to calculate h.

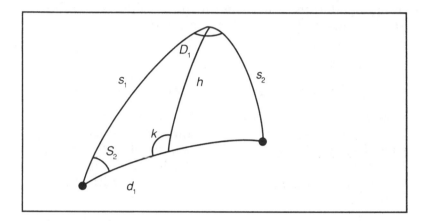

Figure 14-29

☐31. Write a program that reads in the longitude and latitude of two points on the surface of the Earth, calculates the distance between them along the great circle course, the initial angle that the course makes with the meridian of point 1, and then calculates the latitudes and longitudes of several points along the course. Use the result from Exercise 30 for the last part. Note that planes flying from cities in the United States to cities in Europe often fly over the Arctic Ocean, because this allows them to take advantage of the shortest possible course.

32. Calculate the distance between each of these pairs of cities along the great circle course. (Use $r = 6375$ km for the radius of the Earth. Use the program from the previous exercise.)

	Latitude	Longitude
London	51°N	0°
New York	41°N	74°W = 286°E
Los Angeles	34°N	118°W = 242°E
Honolulu	21°N	158°W = 202°E
Anchorage	61°N	150°W = 210°E
Tokyo	36°N	140°E
Sydney	34°S	151°E
Paris	49°N	2°E

□33.   The declination of the sun (in degrees) on a particular date n is approximately

$$d = -23.45\cos[360(n+10)/365]°$$

(where $n = 1$ on January 1, $n = 2$ on January 2, and so on until $n = 365$ on December 31. Ignore leap years and daylight savings time.) The hour angle of the sun at a particular time is approximately

$$H = 360(T - 12)/24°$$

where T is the time measured by a 24 hour clock (0 = midnight, 12 = noon). Write a program that reads in the observer's latitude, and then prints a table giving the altitude angle of the sun for 6am, 9am, 12noon, 3pm, and 6pm for each day of the year.

34.   This formula gives the altitude of a star:

$$\sin(\text{altitude}) = \sin\phi\sin d + \cos\phi\cos d\cos(t - RA)$$

where ϕ = observer's latitude, d = declination of object, RA = right ascension of object, and t = local sidereal time. (See page 248). For given values of ϕ, RA, and d, find a formula for the values of t when the altitude of the star is zero (in other words, it is even with the horizon).

35.   The time between the rising and setting of a star can be approximated by this formula:

$$T = 2\times12/\pi[\pi - \arccos(\tan\phi\tan d)]$$

where ϕ is the latitude of the observer and d is the declination of the star. (All angles are measured in radians. This formula follows from the result of the previous exercise.)

(a)   What happens if $\phi = 0$?

(b)   What happens if $d = 0$?

(c)   What happens if ϕ and d have the same sign?

(d)   What happens if ϕ and d have opposite signs?

(e)   What happens if $\phi + d = \pi/2$?

(f)   What happens if $\pi > (\phi + d) > \pi/2$? (Assume ϕ and d are both positive).

(g)   What happens if $d = \phi - \pi/2$? (Assume $\phi > 0$.)

36.   (a)   Write a program that reads in the observer's latitude, the local sidereal time, the right ascension and declination of a star, and then calculates the altitude of that star.

(b)   For an observer at latitude 40° at sidereal time 10 hours = 150° (corresponding to late evening in March), find the altitude of these stars:

	Right ascension	Declination
Sirius (brightest star)	6h 44m = 101.0°	−16.7°
Alpha Centauri (closest star)	14h 38m = 219.5°	−60.7°
Vega	18h 36m = 279.0°	38.8°
Rigel (in Orion)	5h 13m = 78.3°	−8.2°
Alioth (in Big Dipper)	12h 53m = 193.3°	56.1°
North Star (Polaris)	2h 3m = 30.8°	89.1°

(Note that the declination of the North Star is not exactly 90°.)

37. Suppose you have observed the star Vega to have an altitude of 80° and the direction angle is $M = 70°$. What is your latitude? (You also know that the hour angle H is an acute angle.)

38. Derive the formula $\cos c = \operatorname{ctn} A \operatorname{ctn} B$ for spherical right triangles. (You may use any of the other formulas for spherical right triangles given in the box on page 231.)

39. For small values of x (measured in radians), these equations are approximately true: $\sin x = x$; $\tan x = x$; $\cos x = 1 - x^2/2$. Rewrite the three sides formula and the formulas for $\sin A$, $\cos A$, and $\tan A$ for spherical right triangles for the case where the three sides a, b, and c are all small.

40. Verify that the law of cosines for angles works for the spherical triangle given in Figure 14-20.

41. Consider this spherical triangle:

$$A = 34°, B = 103.156°, C = 61.497°,$$

$$a = 32.66°, b = 70°, c = 58°$$

(In this case the sides have been expressed in degrees.) Suppose you tried to form a new spherical triangle whose angles were equal in measure to the sides of the original triangle, and whose sides were equal in measure to the angles of the original triangle. That would not work, but you can form a spherical triangle whose sides are equal to 180° minus the angles of the original triangle, and whose angles are equal to 180° minus the sides of the original triangle. Such a triangle is called the *polar triangle* of the original triangle.

(a) Calculate the sides and angles for the polar triangle for the triangle given above. (Call the sides a', b', and c'; and call the angles A', B', and C'.)

(b) Verify that the triangle from part (a) satisfies the law of sines for spherical triangles and the law of cosines for sides.

(c)   What is the polar triangle to the polar triangle from part (a)?

42.   Prove the law of cosines for angles. Assume that the law of cosines for sides has already been proven, and then refer to the polar triangle.

★43.   Start with this equation:

$$1 - \cos A = \frac{\sin b \sin c \div \cos b \cos c - \cos a}{\sin b \sin c}$$

(which comes from the law of cosines for sides) and then show that this equation is true:

$$\sin(A/2) = \sqrt{\frac{\sin(s-b)\sin(s-c)}{\sin b \sin c}}$$

where $s = (a + b + c)/2$. This is called a half-angle formula.

★44.   From the equation in the previous exercise it is possible to derive this formula:

$$\cos(A/2) = \sqrt{\frac{\sin s \sin(s-a)}{\sin b \sin c}}$$

Find an expression for $\sin(A/2)\cos(B/2) - \sin(B/2)\cos(A/2)$ and then show that this equation is true:

$$\frac{\sin[(A-B)/2]}{\cos(C/2)} = \frac{\sin[(a-b)/2]}{\sin(c/2)}$$

This equation, and some similar ones, are known as Gauss' or Delambre's analogies.

★45.   Use the equation from the previous exercise, along with this equation:

$$\frac{\cos[(A-B)/2]}{\sin(C/2)} = \frac{\sin[(a+b)/2]}{\sin(c/2)}$$

to derive this equation:

$$\frac{\tan[(A-B)/2]}{\text{ctn}(C/2)} = \frac{\sin[(a-b)/2]}{\sin[(a+b)/2]}$$

This equation, and similar ones, are known as Napier's analogies.

15
Polynomial Approximation for sin x and cos x

Recordis decided that Trigonometeris was very good company during the occasional moments when he was not talking about trigonometry. However, Recordis still was bothered that there was no algebraic expression for the trigonometric functions. "I would sleep much more easily at night if I knew a formula for $\sin x$," he said. "Algebra is a much more manageable subject—it doesn't rely on mysterious functions."

Trigonometeris just laughed. "You will never find an algebraic representation for trigonometric functions! They are too special."

Recordis was determined to try. He made a list of all the complicated algebraic expressions he could think of, but none of them resembled the trigonometric functions. Finally he turned to his favorite type of expressions, polynomials.

★The Quest for the Elusive Algebraic Expression

"There has to be a way to find a polynomial that represents $\sin x$," Recordis claimed. He made a list of properties of polynomials.

A second-degree polynomial curve has one change of direction.

A third-degree polynomial curve can have two changes of direction.

A fourth-degree polynomial curve can have three changes of direction.

A fifth-degree polynomial curve can have four changes of direction.

"But the graph of the function $y = \sin x$ changes directions an infinite number of times!" Trigonometeris reminded him. "You will never find a polynomial to represent it!"

"I'll use a polynomial of infinite degree if I have to!" Recordis cried.

"There is no such thing!" Trigonometeris said.

However, the professor had been listening to the conversation and suddenly became interested. She wrote what she though an infinite-degree polynomial must look like:

$$a_0 + a_1x + a_2x^2 + a_3x^3 + a_4x^4 + a_5x^5 + \ldots$$

"That doesn't work!" Trigonometeris said. "If you add together an infinite number of terms, then the sum will be infinity. We know that the value of $\sin x$ is always less than 1."

"But the sum of an infinite number of terms does not always have to be infinity," the professor told him. "when we were studying algebra we found that the sum of an infinite geometric series can be a finite number. For example,

$$1 + z + z^2 + z^3 + z^4 + z^5 + \ldots$$

"is equal to

$$\frac{1}{1-z}$$

"if $|z| < 1$."

"This is an old trick, but it just might work!" Recordis said. "I bet $\sin x$ really can be represented by an infinite polynomial! And I bet that for most practical purposes we can find an approximate value for $\sin x$ just by taking the first few terms of the infinite polynomial

$$\sin x = a_0 + a_1x + a_2x^2 + a_3x^3 + a_4x^4 + a_5x^5 + a_6x^6 + a_7x^7$$

"Now all we have to do is find a formula to tell us what the coefficients should be." Recordis thought a moment. "In fact, I just thought of one clue. If $x = 0$, then we can ignore all the terms with an x and the formula says $\sin x = a_0$. Since we know that $\sin 0 = 0$, it follows that $a_0 = 0$. Therefore,

$$\sin x = a_1 x + a_2 x^2 + a_3 x^3 + a_4 x^4 + a_5 x^5 + a_6 x^6 + a_7 x^7$$

The professor stared at this expression and came up with a shrewd idea. "We know that the sine function satisfies one very important property:

$$\sin(-x) = -\sin x$$

Suppose we have any polynomial that contains only *odd* powers of x; for example, $f(x) = 10x^7 + 7x^5 - x^3 + 23.4x$. This polynomial will satisfy the same property:

$$f(-x) = -f(x)$$

However, any polynomial that contains an even power of x, such as $g(x) = 12x^3 + 4x^4$, will not satisfy the property

$$g(-x) = -g(x)$$

Therefore, I suggest that $\sin x$ must be represented by a polynomial that contains only odd powers of x:

$$\sin x = a_1 x + a_3 x^3 + a_5 x^5 + a_7 x^7$$

Trigonometeris still thought the others were trying to insult the trigonometric functions by representing them as polynomials, but Recordis and the professor enthusiastically set about finding formulas for a_1, a_3, a_5, and a_7. Since they had four unknowns, they knew they needed four equations to solve for them. Since the equation should be true for any value of x, they picked four values of x: $x = 1$, $x = 0.5$, $x = 1.5$, and $x = 2$. They used the values $\sin 1 = 0.84147$, $\sin 0.5 = 0.47943$, $\sin 1.5 = 0.99479$, and $\sin 2 = 0.90930$. These values gave them four equations:

$$0.84147 = a_1 + a_3 + a_5 + a_7$$
$$0.47943 = 0.5a_1 + 0.125a_3 + 0.03125a_5 + 0.0078125a_7$$
$$0.99749 = 1.5a_1 + 3.375a_3 + 7.59375a_5 + 17.08593a_7$$
$$0.90930 = 2a_1 + 8a_3 + 32a_5 + 128a_7$$

"That is a regular four-equation linear system with four unknowns!" the professor said enthusiastically. Recordis was not quite so enthusiastic, because he knew that solving this type of system was a lot of hard work. However, he soon came up with some results: $a_1 = 1$, $a_3 = -0.16667 = -\frac{1}{6}$, $a_5 = 0.0082986 = \frac{1}{120}$, and $a_1 = -0.0001798$. Therefore,

$$\sin x = x - \frac{x^3}{6} + \frac{x^5}{120} - \cdots$$

★The Factorial Function

"Do you see any pattern that might tell us what the rest of the coefficients are?" the professor asked.

Recordis searched through pages of tables in his notebook. Suddenly he found something interesting under the heading "Factorial Function."

The factorial of a whole number n (symbolized with an exclamation point: $n!$) is equal to

$$n! = n(n-1)(n-2)(n-3)\cdots 5 \times 4 \times 3 \times 2 \times 1$$

For example,

$$1! = 1$$
$$2! = 2$$
$$3! = 6$$
$$4! = 24$$
$$5! = 120$$

"Look!" Recordis said, "6 is 3!, and 120 is 5!. I bet the next term will be $-x^7/7!$, and I bet the term after that will be $x^9/9!$, and then $-x^{11}/11!$, and so on."

We tried many more examples, and this formula seemed to work each time. It still took a lot of convincing before Trigonometeris was willing to accept this result, but he finally began to appreciate the value of an algebraic formula for calculating the sine function. We also found a similar formula for the cosine function. Since $\cos(-x) = \cos x$, the professor guessed that the polynomial representing $\cos x$ must contain only even powers of x. Finally, after much investigation, the king was able to decree:

★Series Representation of sin x and cos x

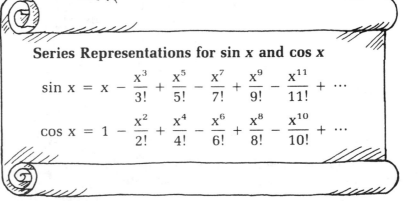

Series Representations for sin x and cos x

$$\sin x = x - \frac{x^3}{3!} + \frac{x^5}{5!} - \frac{x^7}{7!} + \frac{x^9}{9!} - \frac{x^{11}}{11!} + \cdots$$

$$\cos x = 1 - \frac{x^2}{2!} + \frac{x^4}{4!} - \frac{x^6}{6!} + \frac{x^8}{8!} - \frac{x^{10}}{10!} + \cdots$$

(Note that the signs for the terms in each series alternate between plus and minus. Also, be sure to use radian measure for x.)

(These series are examples of a more general type of series called a *Taylor series*. Derivations of Taylor series require the calculus.)

1. How can you approximate $\sin x$ when x is small?

2. How can you approximate $\cos x$ when x is small?

3. Make a table that shows the series approximation for $\sin x$ after two terms of the series, three terms, four terms, and five terms for these values of x: $x = 0.1$; $x = 1$; $x = \pi/2$; $x = \pi/6$, and $x = \pi/4$.

4. Here is a series expression for $\arctan x$:

$$\arctan x = x - \frac{x^3}{3} + \frac{x^5}{5} - \frac{x^7}{7} + \frac{x^9}{9} - \cdots$$

Use this expression to derive a series expression for π.

Answers to Exercises

Chapter 1

1. Use the Pythagorean theorem: $\sqrt{5^2 - 3^2} = 4$

2. 10   5. $\sqrt{2}$ 7. 50

3. 5   6. $\sqrt{10}$ 8. 0.521

4. 25

9. Two complementary angles add to 90°. $90° - 45° = 45°$.

10. 60°   12. 15°   14. 67.5°

11. 30°   13. 0°

15. The sum of the angles in any triangle is 180°.

$$180° - (45° + 45°) = 90°$$

16. 60°   20. 70°   23. 45°

17. 60°   21. 40°   24. 20°

18. 160°   22. 50°   25. 70°

19. 10°

26. 360° (A quadrilateral can be broken up into two triangles.)

27. 540°   28. (n = 2)180°

29. Consider triangle *ABC*. (See Figure A-1.) Draw a line parallel to side *BC* that passes through point *A*. Then, angle 4 + angle 3 + angle 5 = 180° because we have drawn a straight line. Also, angle 1 = angle 4 and angle 2 = angle 5, because these pairs of angles are both alternate interior angles between two parallel lines. (See a book on geometry.) Then, by substitution: angle 1 + angle 2 + angle 3 = 180°.

30. Consider a triangle *ABC*. (See Figure A-2.) Draw a line perpendicular to side *BC* that passes through point *A*. Then, triangle *BAD* is congruent to triangle *CAD* because (1) angle 2 = angle 3 (they are both right angles); (2) side *AD* in triangle *BAD* equals side *AD* in triangle *CAD* (they are, in fact, the same side); and (3) angle 1 = angle 4 (because it is given that this is an isosceles triangle). Then, the length of side *BA*

Figure A-1

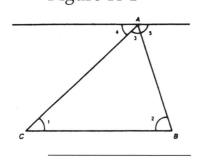

Figure A-2

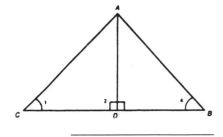

equals the length of *CA* because they are corresponding sides in two congruent triangles.

31. 16° 30 minutes	35. 12.25°
32. 22° 20 minutes	36. 34.83°
33. 8 seconds	37. $0.0011 = 1.1 \times 10^{-3}$
34. 12 minutes 7.2 seconds	38. 5.235°

Chapter 2

	Angle of interest	Adjacent side	Opposite side	Hypotenuse
1.	45°	16	16	$16\sqrt{2}$
2.	45°	2	2	$\sqrt{8}$
3.	45°	1	1	$\sqrt{2}$
4.	40°	10	8.39	13.05
5.	40°	19.66	16.5	25.67
6.	40°	11.11	9.32	14.5
7.	40°	20	16.782	26.11
8.	10°	16.54	2.92	16.80
9.	10°	0.1750	0.0309	0.1777
10.	10°	100	17.633	101.54
11.	10°	567.13	100	575.88
12.	(See Figure A-3.)			

Figure A-3

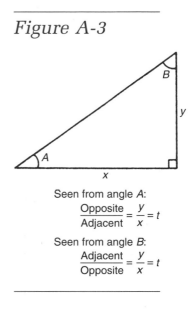

Seen from angle *A*:

$$\frac{\text{Opposite}}{\text{Adjacent}} = \frac{y}{x} = t$$

Seen from angle *B*:

$$\frac{\text{Adjacent}}{\text{Opposite}} = \frac{y}{x} = t$$

Chapter 3

2. Form a right triangle with two 45° angles, with two legs of length 1. Then the hypotenuse has length $\sqrt{2}$. From this information we can calculate sin 45°, cos 45°, and tan 45°.

For a 30-60-90 triangle, if the shortest leg has length 1 then the longest leg has length $\sqrt{3}$ and the hypotenuse has length 2. From this information we can calculate the values of the trigonometric functions for 30 and 60°.

Here is the table you should memorize:

A	$\sin A$	$\cos A$	$\tan A$
30°	$\frac{1}{2}$	$\sqrt{3}/2$	$1/\sqrt{3}$
45°	$1/\sqrt{2}$	$1/\sqrt{2}$	1
60°	$\sqrt{3}/2$	$\frac{1}{2}$	$\sqrt{3}$

3. $\sin A = \text{(opposite side)/hypotenuse}$ and $\cos A = \text{(adjacent side)/hypotenuse}$. Therefore:

$$\frac{\sin A}{\cos A} = \frac{\text{(opposite side)/hypotenuse}}{\text{(adjacent side)/hypotenuse}} = \tan A$$

	A	$\sin A$	$\cos A$	$\tan A$
4.	10°	0.17365	0.98481	0.17633
5.	15°	0.25882	0.96593	0.26795
6.	33.4°	0.55048	0.83485	0.65938
7.	76.6°	0.97278	0.23175	4.19756
8.	16.4°	0.28234	0.95931	0.29432
9.	45°	0.70711	0.70711	1
10.	12°	0.20791	0.97815	0.21256

Figure A-4

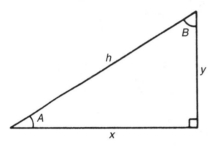

$$\sin A = \frac{y}{h}$$

$$\cos B = \frac{y}{h}$$

$$B = 90° - A$$

$$\cos(90° - A) = \sin A$$

11. See Figure A-4.

12. Since $\sin A = \cos(90° - A)$, we could find the sines of angles greater than 45° by looking at the cosines of their complements. For example, $\sin 75° = \cos(90° - 75°) = \cos 15°$, so we could find the value of $\sin 75°$ by calculating $\cos 15°$. Suppose we needed to find the tangent of an angle greater than 45°. Then we can do this:

$$\tan A = \frac{\sin A}{\cos A} = \frac{\cos(90° - A)}{\sin(90° - A)} = \frac{1}{\tan(90° - A)}$$

For example, to find $\tan 75°$, we could calculate $1/\tan(90° - 15°) = 1/\tan 15° = 1/0.26795 = 3.7320$.

	Angle of interest	Adjacent side	Opposite side	Hypotenuse
13.	30°	50	$50/\sqrt{3}$	$100/\sqrt{3}$
14.	30°	48	$16\sqrt{3}$	$32\sqrt{3}$
15.	30°	$9\sqrt{3}$	9	18
16.	60°	12	$12\sqrt{3}$	24
17.	60°	$8\sqrt{3}$	24	$16\sqrt{3}$
18.	60°	$\frac{1}{2}$	$\sqrt{3}/2$	1
19.	10°	16.34	2.88	16.59
20.	10°	20.64	3.64	20.96
21.	35°	12.98	9.09	15.846
22.	42°	7.3	6.57	9.82
23.	47.5°	10	10.913	14.80
24.	58.4°	16.51	26.84	31.508

25.	45	56.00	56.00	79.20
26.	45	125.80	125.80	177.91
27.	45	606.70	606.70	858.00
28.	2	18.00	0.63	18.01
29.	2	1045.22	36.50	1045.86
30.	2	15.72	0.55	15.73
31.	87	158.00	3014.75	3018.88
32.	87	41.30	788.00	789.08

33. 36.4

34. 274.7

35. 54.5

36. 519.6

37. 8.0

38. 3359

39. 6.3

40. 1389

41. 1.0

42. 45.8

43. 1513

44. 56.86

45. $d = s/[2\tan(A/2)]$

46. 60 miles

47. 8000 feet

48. 384,000 kilometers

49. 384,000 kilometers

50. 93 million miles

51. 800 million miles

52. 2.3×10^{15} miles

53. 2.2 million light years

54. $t = h (\tan A/\tan B - 1)$

55. Let x be the distance from the base of the cliff to the first point. Then $h/x = \tan A_1$ and $h/(x + d) = \tan A_2$. From these equations we can derive

$$h = \frac{d \tan A_1 \tan A_2}{\tan A_1 - \tan A_2}$$

For Exercises 56 to 61, use the formula from Exercise 55 for the top and bottom of the tower.

56. 5.7

57. 4.4

58. 20.7

59. 18.2

60. 267

61. 4.3

62. Set up a right triangle with a 40° angle, an adjacent side of length 1, and an opposite side of length 0.8391.

63. Set up a right triangle with a 10° angle, a hypotenuse of 1, and an opposite side of length 0.1736.

64. Set up the appropriate right triangle, and then look at matters from the point of view of the other angle.

Chapter 4

1. 2.59 north; 9.66 east

2. 17 south; 29.44 east

3. 1.07 north; 4.88 west

4. 0.999 north; 0.052 east

5. 33.55 south; 49.74 west

6. 58.47 north; 191.26 west

7. 56.57 north; 56.57 west

8. Let w be the speed of the wind and v be the airspeed of the plane. Then $\tan A = w/v$, where A is the angle by which the plane is off course. Therefore: $w = v \tan A$. $50 \tan 45 = 50$.

9.	36.4	12.	30.1
10.	178.3	13.	17.0
11.	52.5	14.	83.6

15. 7.82. Use the formula $d = (v_0/g)(2 \sin A \cos A)$.

16.	125	20.	1.28
17.	349	21.	1.33
18.	274	22.	1.73
19.	177	23.	0.56

24. 1.43. (This situation is the same as if the book were dropped freely).

25. The book will not begin to slide if the table is not tilted at all.

26. We can calculate the downfield component (d) of the player's motion from the formula $d = v \cos A$, where A is the angle that the player's course makes with the sideline and v is his total speed (which is 7 yards per second in this case). The time that it takes to run 10 yards will be $10/d$.

Straight downfield: $10/(7 \cos 0°) = 1.429$			
10°	1.451	40°	1.865
20°	1.520	50°	2.222
30°	1.650	60°	2.857

27. Use the formula $\sin A_2 = n_1 \sin A_1/n_2$:

$$\frac{(1)(\sin 41.68°)}{1.33} = 0.5 \quad A_2 = 30°$$

28. $\sin A_2 = 0$ $A = 0$

29. $\sin A_2 = 0.7071$ $A_2 = 45°$

30. $\sin A_2 = 0.2072$ $A_2 = 12°$

31. $\sin A_2 = 0.3178$ $A_2 = 19°$

32. $\sin A_2 = 0.3759$ $A = 22°$

33. $\sin A_2 = 0.5317$ $A = 32°$

34. $\sin A_2 = 0.5760$ $A = 35°$

35. $\sin A_1 = n_2 \sin A_2 / n_1 = 1.33 \sin 40° = 0.8549$
$\qquad A_1 = 59°$

36. In this case $\sin A_1 = 1$, so $A_1 = 90$.

37. If you tried to calculate $\sin A_1$ from Snell's law, you would get a value greater than 1, which is impossible. In this case no light will be refracted from the water to the air. Instead, all the light will be reflected back into the water.

38. We know $n_1 \sin A_1 = n_2 \sin A_2$. In the case n_2 is the index of refraction for the glass. We know $n_1 = 1$, so

$$n_2 = \frac{\sin A_1}{\sin A_2} = \frac{\sin 20°}{\sin 12.34°} = 1.6$$

Chapter 5

1.	60°	15.	208.84°	29.	0.00029
2.	30°	16.	113.85°	30.	4.85×10^{-6}
3.	45°	17.	$\pi/6$	31.	0.0908
4.	36°	18.	$\pi/4$	32.	0
5.	18°	19.	$3\pi/2$	33.	0
6.	15°	20.	1.745	34.	1.6π
7.	72°	21.	3.770	35.	0
8.	57.30°	22.	0.079	36.	$3\pi/2$
9.	114.59°	23.	0.0174	37.	$\pi/2$
10.	171.89°	24.	0.995	38.	0.45π
11.	229.18°	25.	1.012	39.	0.45π
12.	286.48°	26.	1.047	40.	$\pi/2$
13.	94.25°	27.	1.396		
14.	171.17°	28.	1.484		

	A in degrees	A in radians	$\sin A$
41.	4°	0.06981	0.06976
42.	3.5°	0.06109	0.06105
43.	3°	0.05236	0.05234
44.	2.5°	0.04363	0.04362
45.	2°	0.034907	0.034899
46.	1.5°	0.026180	0.026177
47.	1°	0.017453	0.017452
48.	0.5°	0.0087266	0.0087265
49.	0.2°	0.0034907	0.0034907
50.	0.1°	0.0017453	0.0017453

51.   If A is in radians, then $\sin A$ is approximately equal to A for small values of A.

	A in degrees	$\sin A$	$\cos A$	$\tan A$
52.	180°	0	-1	0
53.	270°	-1	0	Undefined
54.	135°	$1/\sqrt{2}$	$-1/\sqrt{2}$	-1
55.	225°	$-1/\sqrt{2}$	$-1/\sqrt{2}$	1
56.	315°	$-1/\sqrt{2}$	$1/\sqrt{2}$	-1
57.	120°	$\sqrt{3}/2$	$-\frac{1}{2}$	$-\sqrt{3}$
58.	150°	$\frac{1}{2}$	$-\sqrt{3}/2$	$-1/\sqrt{3}$
59.	210°	$-\frac{1}{2}$	$-\sqrt{3}/2$	$1/\sqrt{3}$
60.	240°	$-\sqrt{3}/2$	$-\frac{1}{2}$	$\sqrt{3}$
61.	300°	$-\sqrt{3}/2$	$\frac{1}{2}$	$-\sqrt{3}$
62.	330°	$-\frac{1}{2}$	$\sqrt{3}/2$	$-1/\sqrt{3}$
63.	57.30°	0.8415	0.5403	1.5574
64.	114.59°	0.9093	-0.4161	-2.1850
65.	171.89°	0.1411	-0.9900	-0.1425
66.	229.18°	-0.7568	-0.6536	1.1578
67.	286.48°	-0.9589	0.2837	-3.3805
68.	343.77°	-0.2794	0.9602	-0.2910
69.	401.07°	0.6570	0.7539	0.8714

70. $3\pi/4$

71. $11\pi/6$

72. $3\pi/2$

73. $2\pi/3$

74. $\pi/6$

75. $5\pi/4$

76. $\pi/2$

77. $11\pi/6$

78. $2\pi/3$

79. $3\pi/4$

80. $7\pi/6$

81. $d = 2r\sin(s/2r)$

82. 5.64 meters per second

83. $v = r\omega$

84. First, calculate the circumference of the orbit: $C = 2\pi r$. Then calculate the velocity: $v = C/p$, where p is the period. Then calculate the angular velocity: $\omega = v/r$.

Mercury: $v = 4.14$ million kilometers per day; $\omega = 0.071$ radians per day

Venus: $v = 3.02$; $\omega = 0.028$

Earth: $v = 2.58$; $\omega = 0.017$

Mars: $v = 2.09$; $\omega = 0.009$

Jupiter: $v = 1.13$; $\omega = 0.0015$

Saturn: $v = 0.83$; $\omega = 0.00058$

85. 6377 (This planet must be Earth.)

86. 2435 (Mercury)

87. 71,390 (Jupiter)

88. 6050 (Venus)

89. $\sec 30° = 2\sqrt{3}/3$; $\csc 30° = 2$; $\operatorname{ctn} 30° = \sqrt{3}$

90. $\sec 45° = \sqrt{2}$; $\csc 45° = \sqrt{2}$; $\operatorname{ctn} 45° = 1$

91. $\sec 60° = 2$; $\csc 60° = 2\sqrt{3}/3$; $\operatorname{ctn} 60° = 1/\sqrt{3}$

92. 1.064

93. 1.428

94. 1.035

95. 1.020

96. 2.633

97. 1.122

98. −1.155

99. 1.004

100. −5.67

101. −$\pi/2$

102. −$7\pi/6$

103. −$\pi/11$

104. −$5\pi/11$

105. −$11\pi/10$

106. −$16\pi/14$

107. Suppose (x, y) are the coordinates of the point along a ray that makes an angle A with the x axis. Then $(x, -y)$ are the coordinates of a point along a ray that makes an angle $-A$ with the x axis. Therefore

$$\sin(-A) = \frac{-y}{r} = -\sin A$$

$$\cos(-A) = \frac{x}{r} = \cos A$$

108. Consider a right triangle with legs of length x and y, and two acute angles A and A'. (We know $A' =$

$90° − A$.) If y is the side opposite angle A, then $\tan A = y/x$. Looking at the triangle from the point of view of angle A', we know $\tan A' = x/y$. Since $\operatorname{ctn} A = \tan(90° = A)$, it follows that

$$\operatorname{ctn} A = \tan A' = \frac{x}{y} = \frac{1}{\tan A}$$

This proof works for right triangles. It is also possible to prove that $\operatorname{ctn} A = 1/\tan A$ in general.

109.

```
1   REM THIS PROGRAM CALCULATES A TABLE OF THE SIN, COS, AND
2   REM TAN FUNCTIONS
10  PI=3.14159
15  PRINT "DEGREES RADIANS SIN COS TAN"
20  FOR D=0 TO 90
30  R=D*PI/180 'CONVERT TO RADIANS BECAUSE MANY COMPUTERS
31             'WILL ONLY CALCULATE TRIGONOMETRIC FUNCTIONS
32             'FOR ANGLES IN RADIAN MEASURE
40  S=SIN(R) : C=COS(R)
50  PRINT USING "#########"; D;
60  PRINT USING "####.####"; R; S; C;
70  IF D<>90 THEN PRINT USING "####.####"; S/C 'REMEMBER TAN(0) IS UNDEFINED
80  NEXT D
90  PRINT : END
```

Chapter 6

1. $\cos A = \sqrt{1 - \sin^2 A}$

    $\tan A = \sin A / \sqrt{1 - \sin^2 A}$

    $\operatorname{ctn} A = \sqrt{1 - \sin^2 A} / \sin A$

    $\sec A = 1 / \sqrt{1 - \sin^2 A}$

    $\csc A = 1 / \sin A$

2. $-\frac{4}{5}$

3. $2\sin A \cos A = -\frac{24}{25}$

4. $2\tan A / (1 - \tan^2 A) = -\frac{24}{7}$

5. $\cos^2 A - \sin^2 A = \frac{7}{25}$

6. $\sin(A/2) = \sqrt{(1 - \cos A)/2} = \sqrt{\dfrac{9}{10}} = \dfrac{3\sqrt{10}}{10}$

7. $\sin A = -\frac{3}{5}$; $\cos A = -\frac{4}{5}$; $\sin B = -\frac{12}{13}$.

    $\sin(A + B) = \left(-\frac{3}{5}\right)\left(-\frac{5}{13}\right) + \left(-\frac{12}{13}\right)\left(-\frac{1}{5}\right) = -\frac{56}{65}$

8. $\cos A = \frac{5}{13}$; $\sin B = \frac{3}{5}$

    $\cos(A + B) = \left(\frac{5}{13}\right)\left(-\frac{1}{5}\right) - \left(\frac{12}{13}\right)\left(\frac{3}{5}\right) = -\frac{56}{65}$

9. Use the half-angle formula:
    $\sin 15° = \sqrt{(1 - \cos 30°)/2} = \sqrt{2 - \sqrt{3}}/2.$

We can also find $\cos 15° = \sqrt{2 + \sqrt{3}}/2$.

10. $\sin 75° = \sin(60° + 15°) = \left[\sqrt{6 + 3\sqrt{3}} + \sqrt{2 - \sqrt{3}}\right]/4$

11. $\sin 7.5 = \dfrac{\sqrt{2 - \sqrt{2 + \sqrt{3}}}}{2}$

12. $-\sqrt{2 - \sqrt{3}}/2$

13. Use the formula

$$\sin A + \sin B = 2\sin \dfrac{A+B}{2}\cos \dfrac{A-B}{2}$$
$$= 2\sin 45° \cos 30°$$
$$= 2\dfrac{1}{\sqrt{2}}\dfrac{\sqrt{3}}{2}$$
$$= \sqrt{\dfrac{3}{2}}$$

14. $\sin(0 - a) = \sin 0 \cos a - \sin a \cos 0 = -\sin a$

15. $\cos(0 - a) = \cos 0 \cos a + \sin a \sin a = \cos a$

16. $\tan(0 - a) = (\tan 0 - \tan a)/(1 - \tan 0 \tan a) = -\tan a$

17. $\cos(\pi/2 - a) = \cos(\pi/2)\cos a + \sin(\pi/2)\sin a = \sin a$

18. $\sin(\pi/2 - a) = \sin(\pi/2)\cos a - \sin a \cos(\pi/2) = \cos a$

19. $\tan(\pi/2 - a) = [\tan(\pi/2) - \tan a]/[1 + \tan(\pi/2)\tan a]$

    This expression is not useful in this form, but we can divide both top and bottom by $\tan(\pi/2)$:

$$\dfrac{1 - (\tan a)/\tan(\pi/2)}{\dfrac{1}{\tan(\pi/2)} + \tan a}$$

The expressions with $\tan(\pi/2)$ in the denominator can be treated as being zero, so therefore $\tan(\pi/2 - a) = 1/\tan a$.

20. $\sec(A + B) = 1/\cos(A + B)$
$$= 1/(\cos A \cos B - \sin A \sin B)$$

$$\dfrac{\dfrac{1}{\cos A \cos B}}{\dfrac{\cos A \cos B}{\cos A \cos B} - \dfrac{\sin A \sin B}{\cos A \cos B}} = \dfrac{\sec A \sec B}{1 - \tan A \tan B}$$

21. $\csc(A + B) = 1/\sin(A + B)$
$$= 1/(\sin A \cos B + \sin B \cos A)$$
$$= \csc A \csc B/(\text{ctn } B + \text{ctn } A)$$

22. $\frac{1}{2}(1 + \cos 2A) = \frac{1}{2}(1 + \cos^2 A - \sin^2 A) = \frac{1}{2}(2\cos^2 A)$

23. $\cos 2B = 1 - 2\sin^2 B$; let $A = 2B$

    $\cos A = 1 - 2\sin^2(A/2)$

    $\sin^2(A/2) = (1 - \cos A)/2$

    $\sin(A/2) = \sqrt{(1 - \cos A)/2}$

24. $\frac{1}{2}[\sin(A+B)+\sin(A-B)]$
$= \frac{1}{2}(\sin A \cos B + \sin B \cos A + \sin A \cos B$
$\quad - \sin B \cos A)$
$= (\sin A)(\cos B)$

25. $2\sin[(A+B)/2]\cos[(A-B)/2]$
$= 2[\sin(A/2)\cos(B/2)+\sin(B/2)\cos(A/2)]$
$\quad \times [\cos(A/2)\cos(B/2)+\sin(A/2)\sin(B/2)]$
$= 2\sin(A/2)\cos(A/2)\cos^2(B/2)$
$\quad + 2\sin(B/2)\cos(B/2)\sin^2(A/2)$
$\quad + 2\sin(B/2)\cos(B/2)\cos^2(A/2)$
$\quad + 2\sin(A/2)\cos(A/2)\sin^2(B/2)$
$= \sin A \cos^2(B/2) + \sin B \sin^2(A/2)$
$\quad + \sin B \cos^2(A/2) + \sin A \sin^2(B/2)$
$= \sin A + \sin B$

26. $\cos A - \cos B = \sin(\pi/2 - A) + \sin(B - \pi/2)$
Apply the last expression to the formula above.

27. $\sec^2 A + \csc^2 A$
$= 1/\cos^2 A + 1/\sin^2 A$
$= \sin^2 A/(\sin^2 A \cos^2 A) + \cos^2 A/(\sin^2 A \cos^2 A)$
$= 1/(\sin^2 A \cos^2 A)$
$= \sec^2 A \csc^2 A$

28. $\sin[(A+B)+C]$
$= \sin(A+B)\cos C + \sin C \cos(A+B)$
$= \sin A \cos B \cos C + \sin B \cos A \cos C$
$\quad + \sin C \cos A \cos B - \sin C \sin A \sin B$

29. $\cos[(A+B)+C]$
$= \cos(A+B)\cos C - \sin(A+B)\sin C$
$= \cos A \cos B \cos C - \sin A \sin B \cos C$
$\quad - \sin A \cos B \sin C - \sin B \cos A \sin C$

30. $\tan(2A) = \tan(A+A) = 2\tan A/(1-\tan^2 A)$

31. $\sin(4A) = \sin[2(2A)] = 2\sin 2A \cos 2A$
$= 2(2\sin A \cos A)(\cos^2 A - \sin^2 A)$
$= \cos A(4\sin A \cos^2 A - 4\sin^3 A)$
$= \cos A[4\sin A(1-\sin^2 A) - 4\sin^3 A]$
$= \cos A(4\sin A - 8\sin^3 A)$

32. $\sin(5A)$

$$= \sin(4A + A) = \sin 4A \cos A + \sin A \cos 4A$$
$$= \cos^2 A(4\sin A - 8\sin^3 A) + \sin A(8\cos^4 A$$
$$\quad -8\cos^2 A + 1)$$
$$= (1 - \sin^2 A)(4\sin A - 8\sin^3 A)$$
$$\quad + \sin A\left[8(1 - \sin^2 A)^2 - 8(1 - \sin^2 A) + 1\right]$$
$$= 5\sin A - 20\sin^3 A + 16\sin^5 A$$

33. $\cos(3A)$

$$= \cos(2A + A) = \cos 2A \cos A - \sin 2A \sin A$$
$$= (2\cos^2 A - 1)\cos A - 2\sin^2 A \cos A$$
$$= 2\cos^3 A - \cos A - 2\sin^2 A \cos A$$
$$= 2\cos^3 A - \cos A - 2(1 - \cos^2 A)(\cos A)$$
$$= 4\cos^3 A - 3\cos A$$

34. $\cos(4A) = 2\cos^2 2A - 1$

$$= 2(2\cos^2 A - 1)^2 - 1$$
$$= 2(4\cos^4 A - 4\cos^2 A + 1) - 1$$
$$= 8\cos^4 A - 8\cos^2 A + 1$$

35. $\sin 3A = 3\sin A - 4\sin^3 A$ (from Exercise 44.)

Then

$$-\sin 3A + 3\sin A = 4\sin^3 A$$

36. $$\sqrt{\frac{1 + \sin A}{1 - \sin A}} = \sqrt{\frac{(1 + \sin A)(1 + \sin A)}{(1 - \sin A)(1 + \sin A)}}$$
$$= \frac{1 + \sin A}{\sqrt{1 - \sin^2 A}}$$
$$= \frac{1 + \sin A}{\cos A} = \sec A + \tan A$$

37. $(\sin A + \cos A)^2 = \sin^2 A + 2\sin A \cos A + \cos^2 A$
$$= \sin 2A + 1$$

38. $\sec^4 A - \sec^2 A = \sec^2 A(\sec^2 A - 1)$
$$= (\tan^2 A + 1)(\tan^2 A)$$
$$= \tan^4 A + \tan^2 A$$

39. $(\sin 2A)/(\sin A) - (\cos 2A)/(\cos A)$
$$= 2\sin A \cos A/\sin A - (2\cos^2 A - 1)/\cos A$$
$$= 2\cos A - 2\cos A + \sec A$$

40. $\sin 3A - \sin A$

$\qquad = 3\sin A - 4\sin^3 A - \sin A$

$\qquad = 2\sin A - 4\sin A(\tfrac{1}{2})(1 - \cos 2A)$

$\qquad = 2\sin A - 2\sin A + 2\sin A\cos 2A$

$\quad (\sin 3A - \sin A)/(\cos 2A)$

$\qquad = 2\sin A\cos 2A/\cos 2A = 2\sin A$

41. $\operatorname{ctn} B \sec B = (\cos A/\sin A)(1/\cos A)$

$\qquad = 1/\sin A = \csc B$

42. $\cos A + \sin A\tan A = \cos A + \sin^2 A/\cos A$

$\qquad\qquad\qquad\qquad = (\cos^2 A + \sin^2 A)/\cos A$

$\qquad\qquad\qquad\qquad = 1/\cos A = \sec A$

43. $(\sin A + \tan A)/(1 + \sec A)$

$\qquad = (\sin A + \tan A)/(1 + 1/\cos A)$

$\qquad = (\sin A + \tan A)/[(\cos A + 1)/\cos A]$

$\qquad = (\cos A\sin A + \cos A\tan A)/(\cos A + 1)$

$\qquad = (\cos A\sin A + \sin A)/(\cos A + 1)$

$\qquad = \sin A(\cos A + 1)/(\cos A + 1) = \sin A$

44. $\sin 3A = \sin(2A + A)$

$\qquad\qquad = \sin 2A\cos A + \cos 2A\sin A$

$\qquad\qquad = 2\sin A\cos^2 A + (\cos^2 A - \sin^2 A)\sin A$

$\qquad\qquad = \cos^2 A(2\sin A + \sin A) - \sin^3 A$

$\qquad\qquad = (1 - \sin^2 A)3\sin A - \sin^3 A$

$\quad \sin 3A = 3\sin A - 4\sin^3 A$

Printout 2

45.
```
10 INPUT "SIN(A) : " ; SINA
20 INPUT "QUADRANT (1, 2, 3, OR 4) : " ; Q
30 COSA=SQR (1 - SINA ^ 2)  'COSINE OF A
40 IF (Q=2) OR (Q=3) THEN COSA= -COSA
50 TANA=SINA/COSA
60 CTNA=COSA/SINA
70 SECA=1/COSA
80 CSCA=1/SINA
```

Chapter 7

1.	6.258	6.	59.18	11.	60°, 60°, 60°
2.	290.93	7.	19.18	12.	90°, 45°, 45°
3.	99.30	8.	88.88	13.	120°, 30°, 30°
4.	31.458	9.	30.17		
5.	17.52	10.	53.09		

14. Use the law of sines. However, you will find that the length of the third side could be either 20 or 40. If you specify two sides of a triangle and one of the angles that is not between those two sides, then you cannot always uniquely determine the parts of the triangle. See Figure A-5.

15. 2.57, 48.62°, 91.38° (In order to calculate the two angles, you need to calculate the inverse of a trigonometric function. We will discuss this topic more in Chapter 10.)

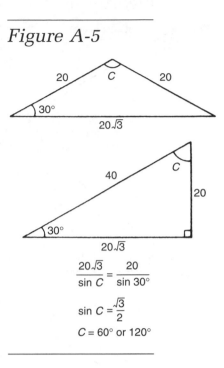

Figure A-5

$$\frac{20\sqrt{3}}{\sin C} = \frac{20}{\sin 30°}$$

$$\sin C = \frac{\sqrt{3}}{2}$$

$$C = 60° \text{ or } 120°$$

16.	3.85, 70.33°, 44.93°	21.	4.08, 4.61, 60°
17.	6.23, 31.73°, 23.23°	22.	2.08, 80°, 80°
18.	2.66, 2.38, 105°	23.	5, 53.13°, 90°
19.	2.57, 3.06, 90°	24.	3.83, 67.5°, 67.5°
20.	2.84, 4.26, 65°		

25. From the law of cosines we can find

$$s = \sqrt{v^2 + w^2 + 2vw\cos(B - A)}$$

26.	618.8	31.	396.5
27.	593.5	32.	502.0
28.	619.9	33.	501.0
29.	404.1	34.	498.5
30.	405.0		

35. $s = \sqrt{v^2 + w^2 + 2vw}$

$\quad = v + w$

36. $s = \sqrt{w^2 + v^2 + 2vw\cos 180°}$

$\quad = \sqrt{w^2 + v^2 - 2vw}$

$\quad = v - w$

37. $s = \sqrt{w^2 + v^2 + 2vw\cos 90°}$

$\quad = \sqrt{w^2 + v^2}$

38. $s^2 = w^2 + v^2 + 2vw\cos(B - A)$

$\quad 0 = w^2 + 2vw\cos(B - A)$

$\quad \cos(B - A) = -w^2/2vw = -w/2v$

39. We originally derived the law of sines by drawing the altitude from side c to angle C. If you draw one of the other altitudes you will see that you can include $c/\sin C$ in the formula.

40–50. Let P be the distance from the planet to the sun, R be the distance from the Earth to the sun, and A

Figure A-6

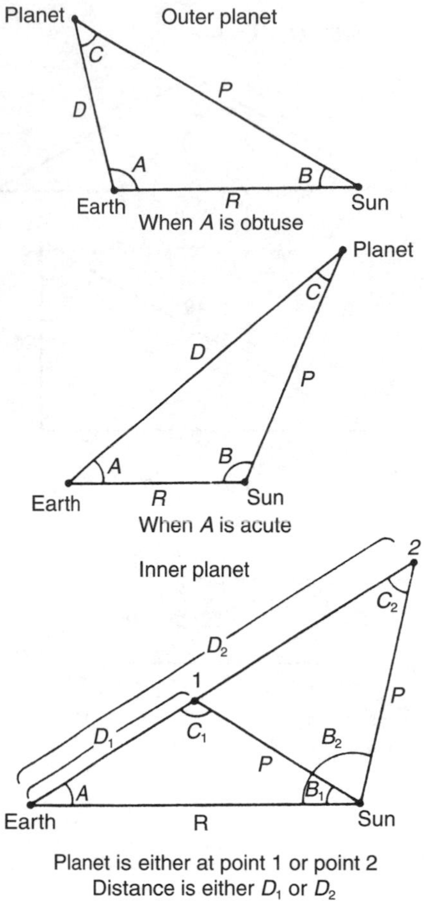

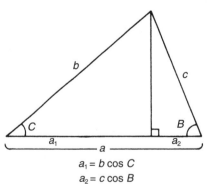

Planet •   Outer planet

C
D
P
A B
Earth R Sun
When A is obtuse

• Planet

C
D
P
A B
Earth R Sun
When A is acute

Inner planet

2
C_2
D_2
1
P
D_1 C_1 B_2
P
A B_1
Earth R Sun

Planet is either at point 1 or point 2
Distance is either D_1 or D_2

Figure A-7

b c
C B
a_1 a_2
a
$a_1 = b\cos C$
$a_2 = c\cos B$

---

be the angle between the sun and the planet as seen from Earth. Define the two angles C and B as shown in Figure A-6. Then

$$\sin C = \frac{R\sin A}{P}$$

Once we have found $\sin C$, we must find the value of C by using the arcsin function on our calculator. Look ahead to Chapter 10 for more information. Then

$$B = 180° - (C + A)$$

$$D = P\frac{\sin B}{\sin A}$$

For Mercury and Venus (the planets that are closer to the sun than Earth) we usually will not be able to unambiguously determine the distance from Earth given this information. This problem is the same problem that is described in Exercise 14. Fortunately, we do not have this problem when we are calculating the distance to an outer planet.

40.   206 or 93   44.   374   48.   635

41.   168 or 114   45.   79   49.   1571

42.   253 or 43   46.   917   50.   1282

43.   164 or 66   47.   763

51.   See Figure A-7.

52.   From the law of sines,

$$\frac{a+b}{c} = \frac{\sin A + \sin B}{\sin C}$$
$$= \frac{2\sin[(A+B)/2]\cos[(A-B)/2]}{2\sin(C/2)\cos(C/2)}$$

Since $A + B + C = 180°$, we know

$$A + B = 180° - C$$
$$\sin\frac{A+B}{2} = \sin\left(90° - \frac{C}{2}\right) = \cos\frac{C}{2}$$

Then

$$\frac{a+b}{c} = \frac{\cos[\frac{1}{2}(A - B)]}{\sin(\frac{1}{2}C)}$$

The proof of the other formula is similar.

53. Use the Mollweide's formulas:

$$\frac{a-b}{a+b} = \frac{\sin[(A-B)/2]\sin(C/2)}{\cos(C/2)\cos[(A-B)/2]}$$

$$= \tan\frac{A-B}{2}\tan\frac{C}{2}$$

$$\tan\frac{C}{2} = \tan\frac{180°-(A+B)}{2}$$

$$= \tan\left(90°-\frac{A+B}{2}\right)$$

$$= \operatorname{ctn}\frac{A+B}{2}$$

$$= \frac{1}{\tan[(A+B)/2]}$$

$$\frac{a-b}{a+b} = \frac{\tan[\frac{1}{2}(A-B)]}{\tan[\frac{1}{2}(A+B)]}$$

54. See Figure A-8. Let A represent the area of the triangle. Then

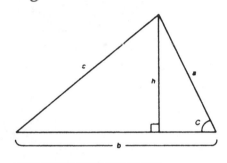

Figure A-8

$$A = \tfrac{1}{2}bh$$

$$= \tfrac{1}{2}ab\sin C$$

$$= \tfrac{1}{2}ab\sqrt{1-\cos^2 C}$$

$$A^2 = \tfrac{1}{4}a^2b^2(1-\cos^2 C)$$

From the law of cosines,

$$\cos C = \frac{a^2+b^2-c^2}{2ab}$$

Then

$$A^2 = \tfrac{1}{4}a^2b^2\left[1 - \frac{(a^2+b^2-c^2)^2}{4a^2b^2}\right]$$

$$= \tfrac{1}{16}\left[4a^2b^2 - (a^2+b^2-c^2)^2\right]$$

$$= \tfrac{1}{16}(2a^2b^2 + 2a^2c^2 + 2b^2c^2 - a^4 - b^4 - c^4)$$

$$= \tfrac{1}{16}(a+b+c)(b+c-a)(c+a-b)(a+b-c)$$

$$= \frac{(a+b+c)}{2}\frac{(b+c-a)}{2}\frac{(c+a-b)}{2}\frac{(a+b-c)}{2}$$

$$A^2 = s(s-a)(s-b)(s-c)$$

$$A = \sqrt{s(s-a)(s-b)(s-c)}$$

55.
```
10 INPUT A, B, C
20 S=(A+B+C)/2
30 AREA=SQR(S*(S-A)*(S-B)*(S-C)) 'AREA OF TRIANGLE FROM HERO'S FORMULA
40 ALT1=2*AREA/A : ALT2=2*AREA/B : ALT3=2*AREA/C
50 REM NOTE THAT THE THREE ALTITUDES CAN BE CALCULATED EASILY
51 REM ONCE THE AREA OF THE TRIANGLE IS KNOWN
```

Printout 3

56.
```
10 INPUT "SIDE A: ";SIDEA
20 INPUT "SIDE B: ";SIDEB
30 INPUT "ANGLE BETWEEN THEM (IN DEGREES): ";C
40 C=C*3.14159/180 'CONVERT TO RADIANS
50 SIDEC=SQR (SIDEA^2+SIDEB^2-2*SIDEA*SIDEB* COS(C))
60 REM USE THE LAW OF COSINES
```

57.
```
10 INPUT "ANGLE A (IN DEGREES): ";A
20 INPUT "ANGLE B (IN DEGREES): ";B
30 INPUT "SIDE BETWEEN THEM: "; SIDEC
40 C=180-(A+B) 'THIRD ANGLE
50 PI=3.14159
60 A=A*PI/180 : B=B*PI/180 : C=C*PI/180 'CONVERT TO RADIANS
70 SIDEA = SIN(A)*SIDEC/SIN(C) 'USE THE LAW OF SINES
80 SIDEB = SIN(B)*SIDEC/SIN(C)
```

58. $D = \sqrt{(50-30)^2 + (80-40)^2} = 44.72.$ $\theta_1 = 19.38°.$ $d_1 = 30.19.$ $c_x = 43.5,$ $c_y = 67.$ The equation of the line is:

$$\frac{y-67}{x-43.5} = -\frac{20}{40}$$

which can also be written

$$y = -0.5x + 88.75$$

    If you know the distances to three transmitters, each pair of transmitters will determine a line, and your exact position can be found by determining the point where these three lines intersect. If you are lost in three-dimensional space, knowing your distance from two transmitters at known locations will define the equation of a plane that contains your position. If the coordinates of the transmitters are (x_1,y_1,z_1) and (x_2,y_2,z_2), the distance between the transmitters is:

$$D = \sqrt{(x_2 - x_1)^2 + (y_2 - y_1)^2 + (z_2 - z_1)^2}$$

Define R_1, R_2, θ_1, d_1, and point C as in Figure 7-10. The coordinates of point C are:

$$\left[c_x = x_1 + \frac{d_1(x_2 - x_1)}{D}, \, c_y = y_1 + \frac{d_1(y_2 - y_1)}{D}, \right.$$

$$\left. c_z = z_1 + \frac{d_1(z_2 - z_1)}{D} \right]$$

The equation of the plane becomes:

$$(x_2 - x_1)x + (y_2 - y_1)y + (z_2 - z_1)z$$
$$= (x_2 - x_1)c_x + (y_2 - y_1)c_y + (z_2 - z_1)c_z$$

If you know the distances to three transmitters, each pair of transmitters defines a plane, and you can determine your exact position by solving for the point where the three planes intersect. This is the basic concept of the global positioning system (GPS), which is a system of 24 satellites orbiting the Earth. A GPS receiver receives signals from three of these satellites. The signals include information about the locations of the satellites, as well as information making it possible to determine the distance to the satellites. (A signal from a fourth satellite is also used to synchronize the time calculation with the receiver.) A computer in the GPS receiver calculates your latitude, longitude, and altitude.

Chapter 8

	Amplitude	Frequency	Angular frequency
1.	9.8	$1/\pi$	2
2.	1	$5/\pi$	10
3.	5	$\frac{1}{2}$	π
4.	π	$\frac{1}{2}$	$1/\pi$
5.	100	$1/200\pi$	$\frac{1}{100}$
6.	4.5	$25/\pi$	50

7. $y = \sin(4\pi x)$

8. $y = \sin(\pi x/8)$

9. $y = \sin(4x)$

10. $y = \sin(8x)$

11–17. See Figure A-9 and Figure A-10.

18. With calculus you may prove that the area under one arch of the sine function is exactly equal to 2.

19.
```
1     REM GENERAL CURVE DRAWING PROGRAM
2     REM THIS PROGRAM WILL DRAW A GRAPH OF
3     REM A CURVE THAT REPRESENTS Y AS A
4     REM FUNCTION OF X. THE USER CAN CHOOSE
5     REM THE SCALE OF THE DRAWING AND THE
6     REM COORDINATES OF THE CENTER OF THE SCREEN.
14    SH = 160 : REM SCREEN HEIGHT
15    S2 = SH/2 : REM HALF OF SCREEN HEIGHT
20    SW = 280 : REM SCREEN WIDTH
30    PRINT "DO YOU WANT AXES SHOWN?"
40    INPUT "Y OR N"; QAS
50    REM SET INITIAL SPECIFICATIONS
60    XC = 0 : YC = 0
61    REM THE CENTER IS AT (0,0)
70    W = 20 : H = 16
71    REM W IS WIDTH, H IS HEIGHT
99    REM ************************
100   REM INPUT SPECIFICATIONS
105   CLS
110   LOCATE 21,1
120   PRINT "W: ";W; " H: ";H;
130   PRINT "CENTER: ";XC; ", ";YC;
140   PRINT " ?"
150   X$=INKEY$: IF X$="" THEN GOTO 150
160   IF X$="G" THEN GOSUB 500 : GOTO 110
170   IF X$="W" THEN GOTO 210
175   IF X$="H" THEN GOTO 220
180   IF X$="X" THEN GOTO 230
```

Printout 6

```
185  IF X$="Y" THEN GOTO 240
187  IF X$="S" THEN END
190  GOTO 150
199  REM ***************************
200  REM CHANGE SPECIFICATIONS
210  INPUT "NEW WIDTH: ";W
215  GOTO 100
220  INPUT "NEW HEIGHT: ";H
225  GOTO 100
230  INPUT "NEW X CENTER: ";XC
235  GOTO 100
240  INPUT "NEW Y CENTER: ";YC
245  GOTO 100
499  REM ***************************
500  REM DRAW DIAGRAM
505  SCREEN 1 : CLS
508  IF QA$ = "Y" THEN GOSUB 700 : REM PLOT AXES
510  XS = XC - W/2 : REM STARTING X VALUE
520  DX = W/SW : REM X INCREMENT
530  X2 = 0
540  FOR X = XS TO (XS+W) STEP DX
550  X2 = X2 + 1
560  GOSUB 1000 : REM CALCULATE Y
570  Y2 = INT(S2 - SH*(Y-YC)/H)
580  PSET (X2,Y2)
590  NEXT X
595  Q$=INKEY$ : IF Q$="" THEN GOTO 595 'PAUSE UNTIL KEY IS PRESSED
600  RETURN
699  REM *********************************
700  REM PLOT AXES
710  XA = SW * (W/2 - XC)/W
715  XA = INT(XA+0.5)
720  YA = SH - (SH * (H/2 - YC)/H)
725  YA = INT (YA + 0.5)
730  IF (XA<1) OR (XA>=SW) THEN GOTO 750
740  LINE (XA, O)-(XA, SH-1)
750  IF (YA<1) OR (YA >= SH) THEN GOTO 770
760  LINE (0,YA)-(SW-1,YA)
770  RETURN
990  REM ******************************
991  REM LINE 1000 DETERMINES THE CURVE
992  REM THAT WILL BE DRAWN. TO CHANGE
993  REM THE CURVE, YOU MUST CHANGE LINE
994  REM 1000. IN THIS EXAMPLE, THE
995  REM CURVE DRAWN IS Y = SIN(X)
1000 Y = SIN(X)
1010 RETURN
```

Chapter 9

	Wavelength	Frequency	Velocity
1.	2π	$1/2\pi$	1
2.	π	$3/2\pi$	$3/2$
3.	1	1	1
4.	5.51	0.55	3.03
5.	π	$1/\pi$	1
6.	0.00079	382	0.3

11.

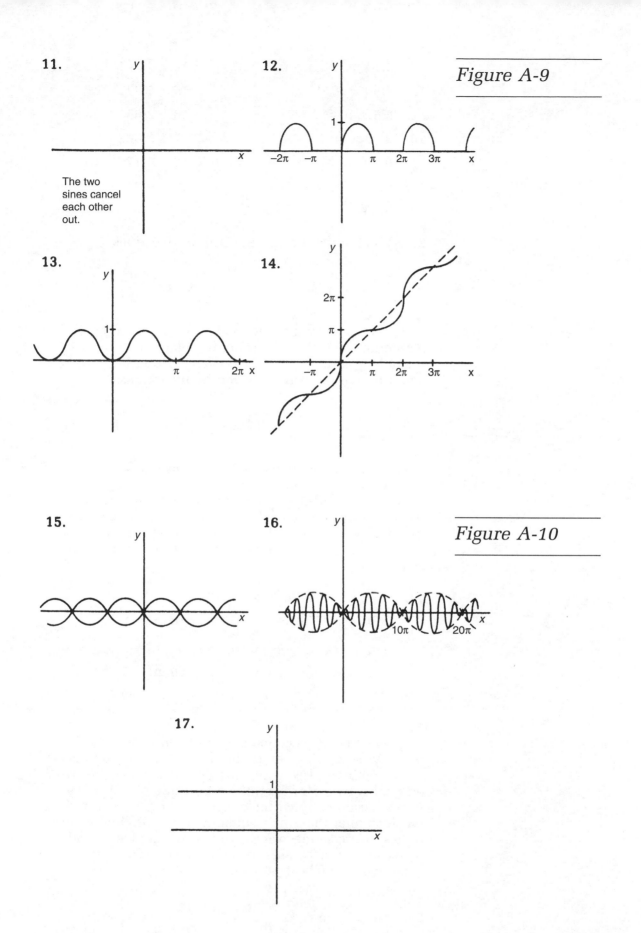

The two sines cancel each other out.

12.

13.

14.

15.

16.

17.

7. 2.14×10^{-12} (in meters)

8. 7.89×10^{-11}

9. 2.5×10^{-7}

10. 3.06×10^{-5}

11. 0.003

12. 0.333

13. 3.26

14. 454.5

15. 1.32

16. 2.65

17. 0.662

18. 0.170

19. 4.24

20. Use the identity for the sum of two sines:

$$y_1 + y_2 = 2A\cos\frac{B}{2}\sin\left(kx - \omega t - \frac{B}{2}\right)$$

We can see that the total wave will also be a harmonic wave with the same wavelength and frequency as the original two waves. However, now the amplitude of the wave is given by the expression

$$2A\cos\frac{B}{2}$$

The amplitude depends on B, the phase difference between the two waves. If $B = 0$, then the two waves are in phase and the amplitude of the total wave is twice the amplitude of the original waves. If $B = \pi$, then the waves are perfectly out of phase and they cancel each other out, meaning that the total wave is zero!

21. $y_1 + y_2 = 2A\sin[(k_1 + k_2)x/2 -$
$(\omega_1 + \omega_2)t/2]\cos[(k_1 - k_2)x/2 - (\omega_1 - \omega_2)t/2]$

Let $k = (k_1 + k_2)/2$, $\omega = (\omega_1 + \omega_2)/2$, $\Delta k = (k_1 - k_2)$, and $\Delta\omega = (\omega_1 - \omega_2)$. Then

$$y_1 + y_2 = 2A\sin(kx - \omega t)\cos\left(\frac{\Delta kx}{2} - \frac{\Delta\omega t}{2}\right)$$

The result is called a *modulated wave*. The underlying wave, given by the factor $\sin(kx - \omega t)$ has a wavelength and frequency almost the same as the two original waves. However, the amplitude of this wave varies periodically with a frequency of $\Delta\omega/4\pi$, which is much smaller than the frequency of the original waves. If these were sound waves, you would notice periodic increases and decreases in the intensity. This situation is known as the beat phenomenon.

22. See Figure A-11.

23. See Figure A-12.

28. $kL = \pi/2 + n\pi$

$k = (\pi/2 + n\pi)/L$

$\lambda = 2\pi L/(\pi/2 + n\pi) = 4\pi L(\pi + 2n\pi) = 4L/(1 + 2n)$

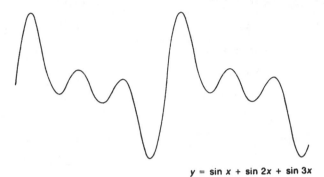

$y = \sin x + \sin 2x + \sin 3x$

Figure A-11

$y = \sin x + 0.5 \sin 2x + 0.25 \sin 4x + 0.125 \sin 8x + 0.062 \sin 16x$

Figure A-12

Chapter 10

1. $-30°$

2. $36°$

3. $-45°$

4. $36.87°$

5. $23.58°$

6. $-17.46°$

7. $22.62°$

8. Use the law of cosines:

$$C = \arccos \frac{a^2 + b^2 - c^2}{2ab}$$

$44.4°, 57.1°, 78.5°$

9. $67.4°, 22.6°, 90.0°$

10. $51.3°, 51.3°, 77.4°$

11. $19.6°, 11.2°, 149.2°$

12. $8.1°, 8.1°, 163.8°$

13. $117.54°, 26.5°, 35.9°$

14. $117.54°, 26.5°, 35.9°$ (Note that the triangle in Exercise 14 is similar to the triangle in Exercise 13.)

15. $39.4°, 54.7°, 85.9°$

16. $36.6°, 63.4°, 16.5$

17. $33.7°, 56.3°, 18.0$

18. $30.5°, 49.5°, 19.4$

19. $11.9°, 18.1°, 24.2$

20. $135°, 15°, 4.3$

21. $18.3°, 61.7°, 13.2$

22. $51.2°, 53.8°, 143.7$

23. $112.5°, 32.5°, 34.1$

24. $52.8°, 62.2°, 20.5$

	Groundspeed	Angle of course
25.	494.1	17.8°
26.	486.1	12.6°
27.	518.0	13.0°
28.	455.1	68.9°
29.	459.4	69.6°
30.	460.0	70.3°
31.	448.4	71.3°
32.	606.2	0.49°
33.	601.4	0.75°
34.	597.3	0.72°
35.	593.1	0.39°

36. Draw right triangles to solve these problems: $\frac{3}{4}$.

37. $\frac{12}{13}$ 41. a

38. $\frac{3}{5}$ 42. $bx/\sqrt{a^2 + b^2 x^2}$

39. $\frac{56}{65}$ 43. 0°, 180°

40. $\sqrt{1 - a^2}$ 44. 60°, 300°

45. There are no solutions to this equation.

46. 0°, 90°

47. $\sin x + \sqrt{1 - \sin^2 x} = \frac{1}{2}$

   After simplifying you will have a quadratic equation in $\sin x$. Use the quadratic formula. You will have to discard one of the roots which is extraneous. The result is

$$\sin x = \frac{1 - \sqrt{7}}{4} \quad x = -24.3°$$

48. Use the identity $\sin^2 x = (1 - \cos 2x)/2$.

$$x = 0°, 180°, 30°, 150°$$

49. Set up a quadratic equation in $\sin^2 x$.

   $x = 30°, -30°, 150°, -150°, 60°, -60°, 120°, -120°$

50. 15°, 165°

51. 0°, 180°, 45°, 135°

52. 60°, 240°, 120°, 300°

53. 0°, 180°

54. −13.29°, −119.55°

55. $1 = \sin(\arcsin 2x + \arcsin x)$

Use the addition rule for the sine function. The final result is $x = 1/\sqrt{5}$.

56.   90°, 270°, 60°, −60°

57.   −15°, −75°, 150°, 210°

58.   45°, 116.57°, 225°, 296.57°

59.   30°, 150°, 270°

60.   45°, 90°, 225°, 270°

61.   210°, 270°, 330°

62.   0°, 180°, 60°, 120°

63.   70.53°, −70.53°

64. $A = 30°, 150°; B = 120°, 240°$

65.
$$\sin(\pi - 2\arctan 2) = \sin(2\arctan 2)$$
$$= 2\sin(\arctan 2)\cos(\arctan 2)$$
$$= 2(2/\sqrt{5})(1/\sqrt{5})$$
$$= \tfrac{4}{5}$$

66.   Calculate $A = \arctan(y/x)$. The result will be correct if the point (x, y) is in the first or fourth quadrants—in other words, if x is positive. If x is negative, then add 180° to $\arctan(y/x)$ to get the final result.

67.   Consider a right triangle with hypotenuse 1, and legs of length y and $\sqrt{1 - y^2}$.

Then: $\sin A = y$, $A = \arcsin y$

$$\tan A = y/\sqrt{1 - y^2}, \quad A = \arctan(y/\sqrt{1 - y^2})$$

Then: $\arcsin y = \arctan(y/\sqrt{1 - y^2})$. Also: $\arccos y = \arctan(\sqrt{1 - y^2}/y)$. (If y is negative, then add π to the result for arccosine, and so the result is between 0 and π.

Chapter 11

	r	θ
1.	22.63	45°
2.	26.93	74.9°
3.	2.24	243.4°
4.	12.08	155.6°
5.	17.46	283.2°
6.	5	126.9°
7.	10	306.9°

8.	13	67.4°
9.	25	253.7°
10.	15.56	315°
11.	15.65	26.6°
12.	24.76	43.4°

	x	y
13.	0	10
14.	5	0
15.	0	−117
16.	−39	0
17.	10.61	10.61
18.	−70.71	70.71
19.	−15.59	9
20.	41.42	17.58
21.	15.88	1.95
22.	23.93	10.16
23.	9.88	−1.56
24.	17.67	−3.43

25–33.   See Figures A-13 to A-15.

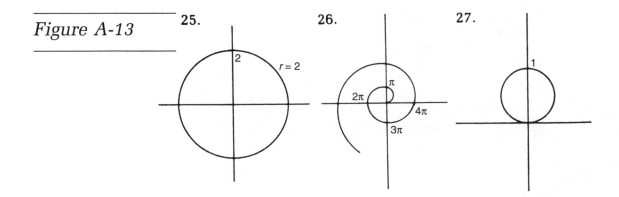

Figure A-13

28. **29.** **30.**

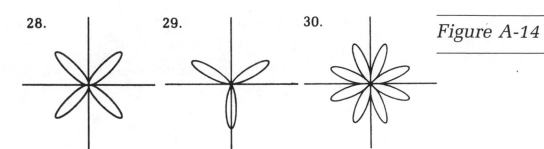

Figure A-14

31. **32.** **33.**

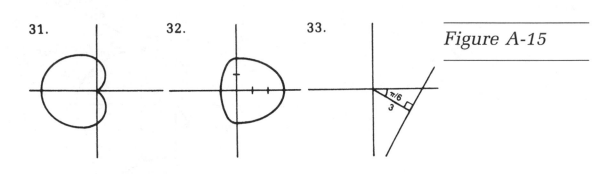

Figure A-15

34. $r^2 = \sec 2\theta$

$r^2 = 1/\cos 2\theta$

$r^2 = 1/(\cos^2 \theta - \sin^2 \theta)$

$r^2 \cos^2 \theta - r^2 \sin^2 \theta = 1$

$x^2 - y^2 = 1$

See Figure A-16.

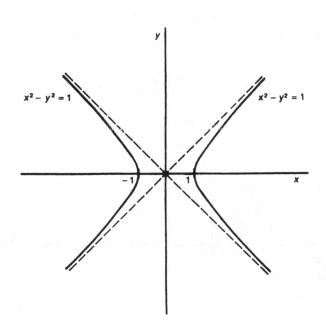

Figure A-16

35. There are four points of intersection:

$$r = 6 \qquad \theta = \pm 60°$$
$$r = 2 \qquad \theta = \pm 120°$$

See Figure A-17.

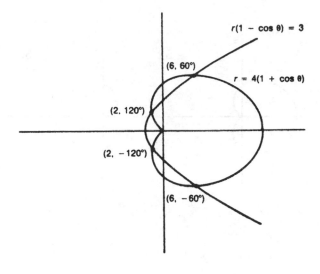

36. $d = r \sin(\theta - \theta_0)$
$d = r \sin\theta \cos\theta_0 - r \sin\theta_0 \cos\theta$
$d = y \cos\theta_0 - x \sin\theta_0$
$y = x \tan\theta_0 + d/\cos\theta_0$

We can see that the slope of the line is $m = \tan\theta_0$ and the y intercept of the line is $d/\cos\theta_0$. See Figure A-18.

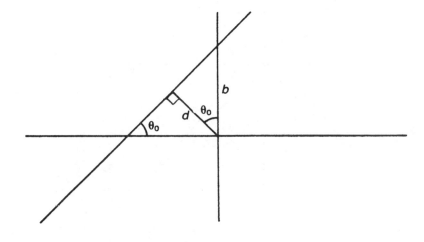

37.
```
1    REM THIS PROGRAM DRAWS A CURVE IN
2    REM POLAR COORDINATES
8    INPUT "N: ";N 'CHANGING N WILL CHANGE THE CURVE
9    INPUT "SCALE: (0 TO 1): ";S: S = S * 90
10   SCREEN 1
20   SH = 200 : SW = 320 'SCREEN HEIGHT AND SCREEN WIDTH
30   FOR T = 1 TO 360
31   REM T IS THE ANGLE MEASURED IN DEGREES
40   RD = 3.14159*T/180
41   REM RD IS THE ANGLE MEASURED IN RADIANS
50   R = S * SIN(N * RD) 'CHANGE THIS STATEMENT TO DRAW
51                        'A DIFFERENT TYPE OF CURVE
59   REM NOW CONVERT TO RECTANGULAR COORDINATES
60   X = (SW/2) + R*COS(RD)
70   Y = (SH/2) - R*SIN(RD)
80   PSET(X,Y) 'LIGHT UP THE POINT AT (X,Y)
90   NEXT T
100  END
```

Chapter 12

	r	θ
1.	5	306.9°
2.	13	202.6°
3.	7.5	90°
4.	11.45	270°
5.	2.83	45°
6.	1	60°
7.	35.9	108.5°
8.	11.7	31.0°
9.	10.8	213.7°
10.	25	73.7°

11. $3 + 4i$

12. $12 + 5i$

13. $7 - 24i$

14. $8 - 6i$

15. $-i$

16. $-0.707 - 0.707i$

17. $-0.866 + 0.5i$

18. $19.28 + 13.25i$

19. $-11.36 - 2.11i$

20. $0.018 + 2.87i$

21. To find the conjugate of a complex number, leave the absolute value unchanged and reverse the sign of the angle.

22. $65(\cos 75.75° + i \sin 75.75°)$

   $(12 + 5i)(3 + 4i) = 16 + 63i$

23. $250(\cos 53.13° + i \sin 53.13°)$

   $(8 + 6i)(24 + 7i) = 150 + 200i$

24. $500(\cos 226.87° + i\sin 226.87°)$

$(5.45 + 49.7i)(-8 + 6i) = (-341.8 - 364.9i)$

25. $4(\cos 210° + i\sin 210°)$

$(1 + 1.732i)(-1.732 + i) = (-3.46 - 2i)$

26. $4(\cos 180° + i\sin 180°)$

$(-1 - 1.732i)(1 - 1.732i) = -4$

27. $\cos 180° + i\sin 180°$

$(1/\sqrt{2} + i/\sqrt{2})(-1/\sqrt{2} + i/\sqrt{2}) = -1$

28. $\cos 270° + i\sin 270°$

$(1/\sqrt{2} + i/\sqrt{2})(-1/\sqrt{2} - i/\sqrt{2}) = -i$

29. $\cos 360° + i\sin 360°$

$(1/\sqrt{2} + i/\sqrt{2})(1/\sqrt{2} - i/\sqrt{2}) = 1$

30. $1552(\cos 85° + i\sin 85°)$

$(8.24 + 13.71i)(87.18 + 42.52i) = 135.3 + 1546i$

31. $3267(\cos 36° + i\sin 36°)$

$(96.46 + 22.27i)(30.38 + 12.89i) = 2643 + 1920i$

32. There are only three cube roots of i. Suppose we try to find an additional third root by writing i in this form:

$$i = \cos 1170° + i\sin 1170°$$

We would find

$$\sqrt[3]{i} = \cos 390° + i\sin 390°$$

But this number can be written

$$\sqrt[3]{i} = \cos 30° + i\sin 30°$$

which is one of the cube roots that we found already.

33. The n nth roots of a complex number with absolute value r will be evenly spaced around a circle of radius $r^{1/n}$.

34. We'll add $r_1(\cos\theta_1 + i\sin\theta_1) + r_2(\cos\theta_2 + i\sin\theta_2)$. Set up a triangle and use the law of cosines and the law of sines. The absolute value of the result will be

$$r = \sqrt{r_1^2 + r_2^2 + 2r_1r_2\cos(\theta_2 - \theta_1)}$$

The angle of the result can be found from either of these two formulas:

$$\theta = \theta_1 + \arcsin\left[r_2\frac{\sin(\theta_2 - \theta_1)}{r}\right]$$

$$\theta = \theta_2 + \arcsin\left[r_1\frac{\sin(\theta_1 - \theta_2)}{r}\right]$$

35. $$\frac{r_1(\cos\theta_1 + i\sin\theta_1)}{r_2(\cos\theta_2 + i\sin\theta_2)}$$

$$= \frac{r_1(\cos\theta_1 + i\sin\theta_2)(\cos\theta_2 - i\sin\theta_2)}{r_2(\cos\theta_2 + i\sin\theta_2)(\cos\theta_2 - i\sin\theta_2)}$$

$$= \frac{r_1(\cos\theta_1\cos\theta_2 - i\sin\theta_2\cos\theta_1 + i\sin\theta_1\cos\theta_2 + \sin\theta_1\sin\theta_2)}{r_2(\cos^2\theta_2 + \sin^2\theta_2)}$$

$$= \frac{r_1[\cos(\theta_1 - \theta_2) + i\sin(\theta_1 - \theta_2)]}{r_2}$$

In words: To divide two polar complex numbers, divide the two absolute values and subtract the angle of the denominator from the angle of the numerator.

36. $(6 + 8i)^4 = [10(\cos 53.13° + i\sin 53.13°)]^4$

$\qquad = 10,000(\cos 212.52° + i\sin 212.52°)$

$\qquad = -8432.0 - 5375.9i$

37. $(24 + 7i)^5 = [25(\cos 16.26° + i\sin 16.26°)]^5$

$\qquad = 9,765,625(\cos 81.3° + i\sin 81.3°)$

$\qquad = 1,477,156 + 9,653,260i$

38. $\cos 135° + i\sin 135°$

39. $\cos 180° + i\sin 180° = -1$

40. $\cos 450° + i\sin 450° = i$

41. $\cos 2835° + i\sin 2835° = 1/\sqrt{2} - 1/\sqrt{2}$

42. $\cos 11.25° + i\sin 11.25°$; $\cos 101.25° + i\sin 101.25°$;

$\cos 191.25° + i\sin 191.25°$; $\cos 281.25° + i\sin 281.25°$

43. $\cos 22.5° + i\sin 22.5°$; $\cos 112.5° + i\sin 112.5°$; $\cos 202.5°$; $+ i\sin 202.5°$; $\cos 292.5° + i\sin 292.5°$

44. $1, i, -1, -i$

45. $1.495(\cos 13.28° + i\sin 13.28°)$; $1.495(\cos 103.28° + i\sin 103.28°)$; $1.495(\cos 193.28° + i\sin 193.28°)$; $1.495(\cos 283.28° + i\sin 282.28°)$

46. $2(\cos 20° + i\sin 20°)$; $2(\cos 110° + i\sin 110°)$; $2(\cos 200° + i\sin 200°)$; $2(\cos 290° + i\sin 290°)$

47. (a) $(1/\sqrt{2})^2 + 2i/2 + (1/\sqrt{2})^2 = 1/2 + i - 1/2 = i$

(b) $(-1/\sqrt{2})^2 + 2i/2 + (-i/\sqrt{2})^2 = 1/2 + i - 1/2 = i$

(c) $(\sqrt{3}/2 + i/2)(\sqrt{3}/2 + i/2)^2$

$\qquad = (\sqrt{3}/2 + i/2)[(\sqrt{3}/2)^2 + 2\sqrt{3}\,i/4 + i^2/4]$

$\qquad = (\sqrt{3}/2 + i/2)(3/4 + \sqrt{3}\,i/2 - 1/4)$

$\qquad = (\sqrt{3}/2 + i/2)(1/2 + \sqrt{3}\,i/2)$

$\qquad = \sqrt{3}/4 + 3i/4 + i/4 + \sqrt{3}\,i^2/4$

$\qquad = i$

Printout 8

48.
```
1     REM THIS PROGRAM CALCULATES THE N DIFFERENT N'TH ROOTS
2     REM OF A COMPLEX NUMBER
5     PI = 3.14159
10    INPUT "REAL PART: ";A
20    INPUT "IMAGINARY PART: ";B
30    INPUT "LEVEL OF ROOTS DESIRED: ";N
40    RO = SQR(A^2 + B^2)
50    THETAO = ATN(B/A)
60    IF A<0 THEN THETAO = THETAO + PI
70    R = RO^(1/N)
80    FOR M = 0 TO (N-1)
90    THETA = (2*PI*M + THETAO)/N
100   A1 = R * COS(THETA) : B1 = R * SIN(THETA)
110   PRINT A1; " + ";B1;"i"
120   NEXT M
130   END
```

Printout 9

49.
```
1     REM THIS PROGRAM CALCULATES (A+BI)^N
5     PI = 3.14159
10    INPUT "REAL PART: ";A
20    INPUT "IMAGINARY PART: ";B
30    INPUT "N: ";N
40    RO = SQR(A^2 + B^2)
50    THETAO = ATN(B/A)
60    IF A<0 THEN THETAO = THETAO + PI
70    R = RO^N
80    THETA = N * THETAO
90    A1 = R * COS(THETA) : B1 = R * SIN(THETA)
100   PRINT A1; " + ";B1; "i"
110   END
```

Chapter 13

	x'	y'
1.	1.414	0
2.	1	−1
3.	0	−1.414
4.	−1	−1
5.	−1.414	0
6.	12.90	12.59
7.	10.39	−6
8.	2.08	−11.82
9.	−6	−10.39
10.	10.57	22.66
11.	−12.50	21.65
12.	71.92	−11.29
13.	9.89	−6.19
14.	73.04	−61.07

15. You may convert the new coordinates (x', y') to the old coordinates (x, y) by rotating the axes by an angle $-\theta$.

16. Let $x' = x + D/2A$; $y' = y + E/2C$. Then the equation becomes

$$A\left(x' - \frac{D}{2A}\right)^2 + C\left(y' - \frac{E}{2C}\right)^2 + D\left(x' - \frac{D}{2A}\right) + E\left(y' - \frac{E}{2C}\right) + F = 0$$

$$Ax'^2 - Dx' + \frac{D^2}{4A} + Cy'^2 - Ey' + \frac{E^2}{4C} + Dx' - \frac{D^2}{2A} + Ey' - \frac{E^2}{2C} + F = 0$$

$$Ax'^2 - \frac{D^2}{4A} + Cy'^2 - \frac{E^2}{4C} + F = 0$$

$$Ax'^2 + Cy'^2 + \left(F - \frac{D^2}{4A} - \frac{E^2}{4C}\right) = 0$$

Let $G = -A/(F - D^2/4A - E^2/4C)$ and let $L = -C/(F - D^2/4A - E^2/4C)$.

17. Let $x' = x + (4CF - E^2)/4CD$; $y' = y + E/2C$. Then the equation becomes

$$C\left(y' - \frac{E}{2C}\right)^2 + D\left(x' - \frac{4CF - E^2}{4CD}\right) + E\left(y' - \frac{E}{2C}\right) + F = 0$$

$$Cy'^2 - Ey' + \frac{E^2}{4C} + Dx' - F + \frac{E^2}{4C} + Ey' - \frac{E^2}{2C} + F = 0$$

$$Cy'^2 + Dx' = 0$$

This is the equation of a parabola with vertex at the origin.

18. Let's let $c = \cos\theta$ and $s = \sin\theta$, where θ is the angle of rotation. Then

$$A' = Ac^2 + Cs^2 + Bcs$$
$$B' = 2cs(C - A) + B(c^2 - s^2)$$
$$C' = As^2 + Cc^2 - Bcs$$

$$B'^2 - 4A'C' = [2cs(C - A) + B(c^2 - s^2)]^2$$
$$- 4[Ac^2 + Cs^2 + Bcs][As^2 + Cc^2 - Bcs]$$
$$= 4c^2s^2(C - A)^2 + 4Bcs(C - A)(c^2 - s^2)$$
$$+ B^2(c^2 - s^2)^2 - 4A^2c^2s^2 - 4ACc^4$$
$$+ 4B^2c^2s^2 - 4ACs^4 - 4C^2c^2s^2$$
$$+ 4ABc^3s - 4ABcs^3$$
$$- 4BCc^3s + 4BCcs^3$$
$$= 4C^2c^2s^2 - 8ACc^2s^2 + 4A^2c^2s^2 +$$
$$4BCc^3s - 4BCcs^3 - 4ABc^3s + 4ABcs^3$$
$$+ B^2c^4 - 2B^2c^2s^2 + B^2s^4 - 4A^2c^2s^2$$
$$- 4ACc^4 + 4B^2c^2s^2 - 4ACs^4$$
$$- 4C^2c^2s^2 + 4ABc^3s - 4ABcs^3$$
$$- 4BCc^3s + 4Bcs^3$$

$$= -8ACc^2s^2 + B^2c^4 + 2B^2c^2s^2 + B^2s^4 -$$
$$\quad 4ACc^4 - 4ACs^4$$
$$= B^2[c^4 + 2c^2s^2 + s^4] -$$
$$\quad 4AC[2s^2c^2 + c^4 + s^4]$$
$$= B^2(c^2 + s^2)^2 - 4AC(c^2 + s^2)^2$$
$$= B^2 - 4AC$$

19.   Let e represent the eccentricity of the conic section. Then

$$\frac{r}{a + r\cos\theta} = e$$

$$r = \frac{ea}{1 - e\cos\theta}$$

20.   Rotate by $-20°$. The new equation becomes

$$16x'^2 + 25y'^2 - 400 = 0$$

$$\frac{x'^2}{25} + \frac{y'^2}{16} = 1$$

The graph will be an ellipse with the axes rotated 20°. See Figure A-19.

Figure A-19

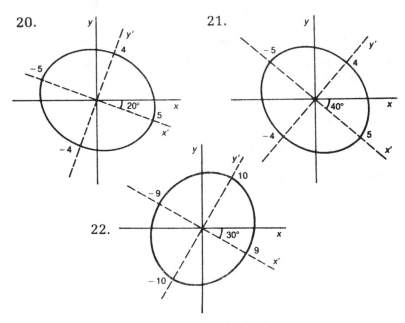

21.   Rotate by $-40°$. The new equation is

$$\frac{x'^2}{25} + \frac{y'^2}{16} = 1$$

22.   Rotate by $-30°$. The new equation is

$$\frac{x'^2}{81} + \frac{y'^2}{100} = 1$$

23. Rotate by −30°. The new equation is

$$\frac{x'^2}{81} + \frac{y'^2}{100} = 1$$

This is the equation of a hyperbola.

24. Rotate by −10°. The new equation is

$$\frac{x'^2}{4} + \frac{y'^2}{9} = 1$$

See Figure A-20.

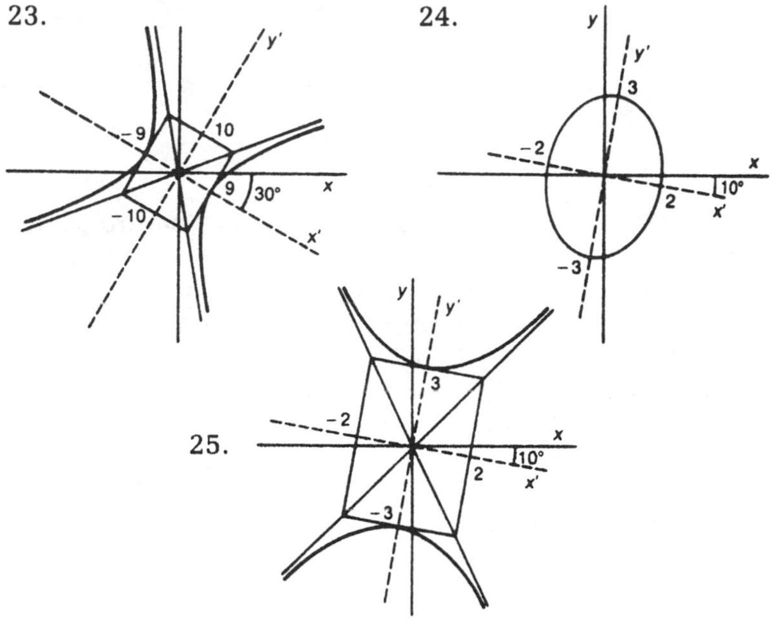

Figure A-20

25. Rotate by −10°. The new equation is

$$\frac{-x'^2}{4} + \frac{y'^2}{9} = 1$$

26. In this equation we have $B = 1$, $F = -1$, and all the other coefficients equal to zero. The angle of rotation is 45°, and the new equation is

$$\frac{x'^2}{2} - \frac{y'^2}{2} = 1$$

This equation fits the standard form for a hyperbola. (Of course, in this case we could have graphed the curve without doing the rotation.) See Figure A-21.

27. In some cases the graph might consist of only a single point (for example, $x^2 + y^2 = 0$). In other cases the graph might consist of two lines (for example, $x^2 - y^2 = 0$). In other cases there will be no points that are solutions to the equation (for example, $x^2 + y^2 + 1 = 0$). These situations are called *degenerate* conic sections.

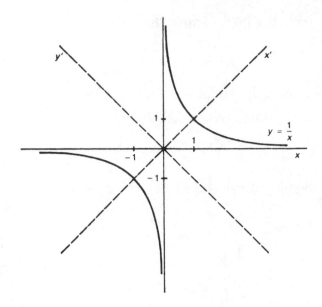

$y = \dfrac{1}{x}$

Calculate $B^2 - 4AC$ to determine the nature of the graph.

	Nature of graph	Angle of rotation
28.	Hyperbola	1.7°
29.	Ellipse	−30.7°
30.	Hyperbola	45°
31.	Ellipse	7.0°
32.	Parabola	−18.4°
33.	Parabola	−26.6°

Printout 10

34.
```
1    REM THIS PROGRAM READS IN THE COEFFICIENTS OF THE EQUATION
2    REM AX^2 + BXY + CY^2 + DX + EY + F = 0
3    REM AND THEN DRAWS THE GRAPH OF THE SOLUTION.
10   PI=3.14159
20   XYSWITCH=0      'IF XYSWITCH IS SET TO 1 THEN IT MEANS THAT
21                   'THE GRAPH NEEDS TO BE DRAWN WITH THE X AND
22                   'Y AXES SWITCHED
100  INPUT "COEFFICIENT OF X^2:   ";A
101  INPUT "COEFFICIENT OF XY:    ";B
102  INPUT "COEFFICIENT OF Y^2:   ";C
103  INPUT "COEFFICIENT OF X:     ";D
104  INPUT "COEFFICIENT OF Y:     ";E
105  INPUT "CONSTANT TERM:        ";F
110  DISC=B^2-4*A*C   'CALCULATE DISCRIMINANT
115  IF ABS(DISC)<.00001 THEN DISC=0
120  IF DISC=0 THEN TYPE=1:PRINT "PARABOLA"
130  IF DISC<0 THEN TYPE=2:PRINT "ELLIPSE"
140  IF DISC>0 THEN TYPE=3:PRINT "HYPERBOLA"
150  IF B=0 THEN THETA=0:GOTO 160
152  IF A=C THEN THETA=PI/4:GOTO 160
154  THETA=1/2 * ATN(B/(A-C)) 'ANGLE OF ROTATION
160  PRINT "ROTATION: ";THETA*180/PI
165  Q$=INKEY$:IF Q$=" " THEN GOTO 165 'PAUSE UNTIL KEY IS PRESSED
170  CS=COS(THETA) : SN=SIN(THETA)
175  GOSUB 2000    ' SET SCREEN FOR GRAPHICS AND DRAW AXES
179  REM CALCULATE COEFFICIENTS OF NEW EQUATION AFTER ROTATION
```

```
180  A1 = A*CS^2 + C*SN^2 + B*SN*CS
190  C1 = A*SN^2 + C*CS^2 - B*SN*CS
200  D1 = D*CS + E*SN
210  E1 = -D*SN + E*CS
220  F1 = F
230  IF TYPE=1 THEN GOSUB 700 ELSE IF TYPE = 2 THEN GOSUB 800
235  IF TYPE=3 THEN GOSUB 900
240  END
699  '
700  REM DRAW GRAPH OF PARABOLA
710  GOSUB 1000    'DETERMINE TRANSLATION
720  FOR X2 = -20 TO 20 STEP .2
730     Y2 = -A1*X2^2/E1
740     GOSUB 2100    'PLOT POINT
750  NEXT X2
760  RETURN
799  '
800  REM DRAW GRAPH OF ELLIPSE OR CIRCLE
810  GOSUB 1100    'DETERMINE TRANSLATION
820  XMAX=SQR(-F2/A1)
830  FOR X2=-XMAX TO XMAX STEP .2
840     Y2 = SQR((-F2-A1*X2^2)/C1)
850     GOSUB 2100 'PLOT POINT
860     Y2 = -Y2
870     GOSUB 2100
880  NEXT X2
890  RETURN
899  '
900  REM DRAW GRAPH OF HYPERBOLA
910  GOSUB 1100    'DETERMINE TRANSLATION
915  IF (A1*F2)>0 THEN XYSWITCH=1:SWAP A1,C1
920  XMIN=SQR(-F2/A1)
930  FOR X3=XMIN+.1 TO XMIN+20 STEP .2
935     X2=X3
940     Y2 = SQR((-F2-A1*X2^2)/C1)
950     GOSUB 2100 'PLOT POINT
960     Y2 = -Y2 : GOSUB 2100
970     X2 = -X3 : GOSUB 2100
980     Y2 = -Y2 : GOSUB 2100
990  NEXT X3
995  RETURN
999  '
1000 REM DETERMINE TRANSLATION FOR PARABOLA
1010 IF A1=0 THEN XYSWITCH=1:SWAP A1,C1:SWAP D1,E1
1020 H=D1/(2*A1):K=(4*A1*F1-D1^2)/(4*A1*E1)
1030 RETURN
1099 '
1100 REM DETERMINE TRANSLATION FOR ELLIPSE OR HYPERBOLA
1110 H=D1/(2*A1) : K=E1/(2*C1)
1115 F2-F1-(D1^2/(4*A1)+E1^2/(4*C1))
1120 RETURN
1999 '
2000 REM SET SCREEN TO GRAPHICS AND DRAW AXES
2010 SCREEN 2
2020 SH=200 : SW=500      'SCREEN HEIGHT AND SCREEN WIDTH
2030 YC=SH/2 : XC=SW/2   'COORDINATES OF CENTER OF SCREEN
2040 LINE(XC,1)-(XC,SH) 'Y AXIS
2045 LINE(1,YC)-(SW,YC) 'X AXIS
2050 XSCALE=50 : YSCALE=50        'THESE CAN BE ADJUSTED TO CHANGE
2051                              'SCALE OF DIAGRAM
2060 RETURN
2099 '
2100 REM PLOT POINT
2110 X4 = X2 : Y4 = Y2
2120 IF XYSWITCH=1 THEN SWAP X4,Y4
2130 X1 = X4 - H : Y1 = Y4 - K 'PERFORM TRANSLATION
2140 X = X1*CS - Y1*SN    'PERFORM ROTATION
2150 Y = Y1*CS + X1*SN
2160 XLOC=SW*X/XSCALE+XC : YLOC=SH-(SH*Y/YSCALE+YC)
2170 PSET(XLOC,YLOC)
2190 RETURN
```

35. $x'^2 + y'^2 + z'^2 =$

$$= x^2\cos^2\theta\cos^2\phi + y^2\sin^2\theta\cos^2\phi + z^2\sin^2\phi$$
$$+2xy\cos\theta\cos^2\phi\sin\theta - 2xz\cos\theta\cos\phi\sin\phi$$
$$-2yz\sin\theta\cos\phi\sin\phi + y^2\cos^2\theta + x^2\sin^2\theta$$
$$-2xy\cos\theta\sin\theta + z^2\cos^2\phi + x^2\cos^2\theta\sin^2\phi$$
$$+y^2\sin^2\theta\sin^2\phi + 2xz\cos\theta\sin\phi\cos\phi$$
$$+2yz\sin\theta\sin\phi\cos\phi + 2xy\cos\theta\sin\theta\sin^2\phi$$

$$= x^2[\cos^2\theta(\sin^2\phi + \cos^2\phi) + \sin^2\theta]$$
$$+y^2[\sin^2\theta(\cos^2\phi + \sin^2\phi) + \cos^2\theta]$$
$$+z^2(\sin^2\phi + \cos^2\phi)$$
$$+2xy(\cos\theta\cos^2\phi\sin\theta - \cos\theta\sin\theta + \cos\theta\sin\theta\sin^2\phi)$$
$$+2xz(\cos\theta\sin\phi\cos\phi - \cos\theta\cos\phi\sin\phi)$$
$$+2yz(\sin\theta\sin\phi\cos\phi - \sin\theta\cos\phi\sin\phi)$$

$$= x^2(\cos^2\theta + \sin^2\theta) + y^2(\sin^2\theta + \cos^2\theta) + z^2$$
$$+2xy(\cos\theta\sin\theta - \cos\theta\sin\theta)$$

$$= x^2 + y^2 + z^2$$

Note that the distance from the point to the origin is the same in either coordinate system since

$$x'^2 + y'^2 + z'^2 = x^2 + y^2 + z^2$$

Chapter 14

1. third side $= \arccos[\cos(0.6) \times \cos(1.4)] = 1.43$
angles: 34.8°; 84.4°

2. third side = 0.6 rad; angles: 50.5°; 50.5°

3. third side = 0.03 rad; angles: 90°; 36.9°

4. third angle = 60.7°; sides: 0.15 rad; 0.30 rad

5. third side $= \arccos[\cos^2(0.7) + \sin^2(0.7)\cos(40°)]$
$= 0.444$ rad angles: 74.4°; 74.4°

6. Use the law of cosines for angles for the angle opposite 0.6093:

$$\cos C = (\cos .6093 - \cos 1.5\cos 1.5)/(\sin 1.5 \sin 1.5)$$

$$C = 35°$$

Find the other two angles with same procedure: 88.7°; 88.7°

7. To find the side opposite 60°:

$$\cos c = (\cos 60° + \cos 69.5° \cos 54.7°)/(\sin 69.5° \sin 54.7°)$$

$$= 0.9188$$

$$c = 0.406; \text{ other sides: } 0.44, \ 0.38$$

8. To find the angle opposite 0.7056:

$$\cos C = -\cos 34.67° \cos 29.69°$$
$$+ \sin 34.67° \sin 29.69° \cos 0.7056$$

$C = 120°$; the two sides measure 0.38 and 0.44.

9. Use the law of sines for the angle opposite 1.2:

$$\sin C = \sin 1.2 \sin 80°/\sin 1.42 = 0.9284$$

C could be either 68.2° or 111.8°; the diagram indicates that it is an acute angle. Use this formula to find the third side:

$$\cos b = \frac{\cos 1.2 \cos 1.42 - \sin 1.2 \sin 1.42 \cos 80° \cos 68.2°}{1 - \sin 1.2 \sin 1.42 \sin 80° \sin 68.2°}$$

$b = 1.6$; other angle = 95.2°

10. The three sides are all $\pi/2$.

11. third side: 0.179; angles: 134° and 33°

12. third angle 91.4°; sides: 0.68 and 0.326

13. third side: 0.3; angles: 40.6°; 104.3°

14. third angle: 94.2°; sides: 2 and 2

15. third side: 0.08; angles: 17.2° and 150°

16. Convert each side to radians (using $r = 6375$).

Here are the calculations for the three-sides rule:

Figure 14-1: $\cos 0.16437 = 0.9865$
$$= \cos 0.1 \cos 0.13067$$
$$= 0.995 \times 0.9915$$

Figure 14-2: $\cos 0.83209 = 0.6733$
$$= \cos 0.78431 \cos 0.31373$$
$$= 0.7079 \times 0.9512$$

17. $\sqrt{r^2\cos^2\phi\cos^2\theta + r^2\cos^2\phi\sin^2\theta + r^2\sin^2\phi}$

$$= r\sqrt{\cos^2\phi(\cos^2\theta + \sin^2\theta) + \sin^2\phi}$$

$$= r\sqrt{\cos^2\phi + \sin^2\phi} = r$$

18. latitude = $\arctan(z/\sqrt{x^2 + y^2})$; longitude = $\arctan(y/x)$ (Make sure that the longitude is in the correct quadrant. See Chapter 10, Exercise 66.)

19. $x = 6375$ $y = 0$ $z = 0$

20. $x = 0$ $y = 6375$ $z = 0$

21. $x = 0$ $y = 0$ $z = 6375$

22. $x = 0$ $y = 5520.9$ $z = 3187.5$

23. $x = 0$ $y = -6375$ $z = 0$

	Latitude	Longitude
24.	−30	250
25.	−60	80
26.	40	120
27.	40	150
28.	70	120

29. (a) $\cos d = \sin^2\phi + \cos^2\phi\cos D$

(d is the distance between the two points in radians)

(b) The radius (r_2) of the circle of latitude whose latitude is equal to ϕ can be found from this formula: $r_2 = r\cos\phi$

Then the distance $= rD\cos\phi$, where D, the difference in longitude, is measured in radians.

(c) The hunter's course is a constant latitude course, not a great circle course, but for a triangle this small, the difference between the two is very small.

(d) distance along great circle course:

$$\cos d = \sin^2(42°) + \cos^2(42°)\cos(100°) = 0.3518$$

$$\text{distance} = r\, \text{arccos}\, 0.3518$$

$$= 6375 \times (1.2113 \text{ rad})$$

$$= 7722 \text{ km}$$

distance along constant latitude course:

$$6375 \times (1.7453 \text{ radians}) \times \cos 42° = 8268 \text{ km}$$

30. From the law of cosines for angles:

$$\cos K = -\cos S_2 \cos D_1 + \sin S_2 \sin D_1 \cos s_1$$

Once K is known, use the law of cosines again:

$$\cos h = (\cos S_2 + \cos D_1 \cos K)/(\sin D_1 \sin K)$$

31 and 32. See this book's web page at *http://www.spu.edu/~ddowning/easytrig.html*.

Printout 11

33.

```
1 REM THIS PROGRAM CALCULATES THE ALTITUDE ANGLE OF THE SUN
10 PI = 3.14159
20 DEF FNASN(X) = ATN(X/SQR(1-X*X))
30 INPUT "ENTER YOUR LATITUDE IN DEGREES: ";LAT
40 LATR = PI*LAT/180      'CONVERT TO RADIANS
100 FOR N = 1 TO 365 : PRINT USING "####";N;
110   DEC = -23.45 * COS((PI/180)*(360*(N+10)/365))
115   DECR = DEC*PI/180  'CONVERT TO RADIANS
120   FOR T = 6 TO 18 STEP 3
130     H = 360 * (T - 12)/24
140     SINALT = SIN(LATR)*SIN(DECR) + COS(LATR)*COS(DECR)*COS(PI*H/180)
150     ALT = 180*FNASN(SINALT)/PI
160     PRINT USING '#######.#";ALT;
170   NEXT T
180   PRINT
190 NEXT N
```

34. $0 = \sin\phi \sin d + \cos\phi\cos d \cos(t - RA)$

$\sin\phi \sin d = -\cos\phi\cos d \cos(t - RA)$

$\cos(t - RA) = -\tan\phi\tan d$

$t - RA = \pm\arccos(\tan\phi\tan d)$

$t = RA \pm \arccos(\tan\phi\tan d)$

35. (a) If $\phi = 0$, then arccos(tan ϕ tan d) is $\pi/2$, and the star will be visible for 12 hours. Therefore, if you live on the equator, all stars will be in the sky for 12 hours, regardless of their location.

(b) If $d = 0$, then the star will again be visible for 12 hours (see the previous part). $d = 0$ for stars along the celestial equator. These stars will be visible for 12 hours, regardless of your latitude.

(c) tan ϕ tan d will be positive, so the arccos will be less tan $\pi/2$, and [π − arccos(tan ϕ tan d) will be greater than $\pi/2$, and the star will be in the sky for more than 12 hours. Note that, if the star is north of the celestial equator, it will stay above the horizon longer the farther north you go. For example, in the spring and summer the sun is north of the celestial equator, and it is above the horizon longer the farther north you are.

(d) The result will be the opposite of the previous part. The star will be above the horizon for less than 12 hours. For example, in the fall and winter the sun is south of the celestial equator, and it is above the horizon for a shorter time the farther north you are.

(e) $\tan(\phi + d) = \infty$

$$\frac{\tan\phi + \tan d}{1 - \tan\phi\tan d} = \infty$$

$$\tan\phi\tan d = 1$$

$$T = 2 \times 12/\pi[\pi - \arccos 1]$$

$$T = 2 \times 12/\pi \times \pi = 24$$

These stars are in the sky for 24 hours—that is, they never set. They will just graze the northern horizon (assuming our observer is in the northern hemisphere). For example, if you are at latitude 40° north, all stars with declination of 50° or greater will be above the horizon all the time.

(f) $\tan(\phi + d) < 0$

which means tan ϕ tan $d > 1$

Then the formula requires you to take the arccosine of a negative number. These stars never touch the horizon at all. See part (e).

$$\phi - d = \pi/2$$

(g) $$\tan(\phi - d) = \infty$$

$$\tan\phi\tan d = -1$$

$$T = 2 \times 12/\pi[\pi - \arccos(-1)]$$

$$T = 2 \times 12/\pi \times (\pi - \pi) = 0$$

These stars just touch the southern horizon at

one time, but they never rise above it. If $d = \phi - K$, where $K > \pi/2$, you would have to take the arccosine of a number less than negative 1. These stars never even touch the horizon. For example, if you are at latitude 40°, you will never see a star at declination −51°.

36.
```
1 REM   THIS PROGRAM READS IN THE LATITUDE OF THE OBSERVER,
2 REM   THE LOCAL SIDEREAL TIME, AND THE RIGHT ASCENSION AND
3 REM   DECLINATION OF A STAR, AND THEN CALCULATES THE ALTITUDE
4 REM   OF THAT STAR
10 PI = 3.14159
20 DEF FNASN(X)=ATN(X/SQR(1-X*X))
30 INPUT   "ENTER LATITUDE IN DEGREES: ";LAT
40 INPUT   "ENTER LOCAL SIDEREAL TIME IN HOURS: ";T
50 INPUT   "ENTER RIGHT ASCENSION OF STAR IN HOURS: ";RA
60 INPUT   "DECLINATION OF STAR IN DEGREES: ";DEC
70 REM     CONVERT EACH MEASUREMENT TO RADIANS
80 LATR=LAT*PI/180
90 TR=T*PI/12
100 RAR=RA*PI/12
110 DECR=DEC*PI/180
120 HR = TR-RAR  'HOUR ANGLE IN RADIANS
130 ALT=FNASN(SIN(LATR)*SIN(DECR)+COS(LATR)*COS(DECR)*COS(HR))
140 ALT=180*ALT/PI  'CONVERT TO DEGREES
150 PRINT   "ALTITUDE: ";:PRINT USING "####.#";ALT
160 END
```

(b) Sirius 17.3°; Alpha Centauri −25.4° (this star is not visible); Vega 1.6°; Rigel 8.4°; Alioth 57.6°; Polaris 39.6°.

37.
$$\sin H = \sin z \sin M / \sin d'$$
$$= \sin 10° \sin 70° / \sin 51.2°$$
$$= 0.20938$$
$$H = 12.086°$$
$$\cos s = \frac{\cos 51.2° \cos 10° - \sin 51.2° \sin 10° \cos 12.086° \cos 70°}{1 - \sin 51.2° \sin 10° \sin 12.086° \sin 70°}$$
$$s = 54°; \text{ latitude} = 36°$$

38. $\operatorname{ctn} A \operatorname{ctn} B = \dfrac{1}{\tan A} \times \dfrac{1}{\tan B}$

$$= \frac{\sin b \sin a}{\tan a \tan b}$$
(using the formulas for tangents on page 231)
$$= \frac{\cos a \sin b \cos b \sin a}{\sin a \sin b}$$
$$= \cos a \cos b$$
$\operatorname{ctn} A \operatorname{ctn} B = \cos C$

(using the three-sides formula)

39. three sides formula:
$$1 - c^2/2 = (1 - a^2/2)(1 - b^2/2)$$
$$1 - c^2/2 = 1 - a^2/2 - b^2/2 + a^2 b^2/4$$

If a and b are very small, then $a^2 b^2$ will be so small that it can be treated as if it were zero. Then the equation

302 *Answers to Exercises*

can be written $c^2 = a^2 + b^2$, which is the same as the Pythagorean theorem:

$$\sin A = a/c$$
$$\cos A = b/c$$
$$\tan A = a/b$$

This exercise shows that spherical triangles whose sides are small (relative to the sphere) behave approximately the same as plane triangles. This matches our everyday experience; we are used to treating a small portion of the surface of the Earth as if it were a plane.

40. From Figure 14-20:

$$-\cos S_1 \cos S_2 + \sin S_1 \sin S_2 \cos d$$
$$= -\cos 61.51° \cos 103.16°$$
$$+ \sin 61.51° \sin 103.16° \cos 0.57$$
$$= -0.4770 \times (-0.2277) + 0.8789 \times 0.9737 \times 0.8419$$
$$= 0.829$$

Since $\cos D = \cos 34° = 0.829$, the law of cosines for angles works.

41. (a) $a' = 146°$; $b' = 76.844°$; $c' = 118.503°$;

$$A' = 147.34°, \; B' = 110°; \; C' = 122°$$

(b) $\sin 146°/\sin 147.34° = \sin 76.844°/\sin 110°$

$$= \sin 118.503°/\sin 122°$$
$$= 1.036$$
$$\cos 118.503° = \cos 146° \cos 76.844°$$
$$+ \sin 146 \sin 76.844°$$
$$\times \cos 122°$$
$$= -0.4772$$

(c) the original triangle

42. Write the law of cosines for sides for the polar triangle from Exercise 41:

$$\cos c' - \cos a' \cos b' + \sin a' \sin b' \cos C'$$

Since $c' = 180 - C$, and so on, we can rewrite the equation in terms of the parts of the original triangle:

$$\cos(180 - C) = \cos(180 - A)\cos(180 - B)$$
$$+ \sin(180 - A)\sin(180 - B)\cos(180 - c)$$

Since $\cos(180 - x) = -\cos x$, and $\sin(180 - x) = \sin x$, we have:

$$-\cos C = (-\cos A)(-\cos B)$$
$$+ \sin A \sin B(-\cos c)$$
$$\cos C = -\cos A \cos B + \sin A \sin B \cos c$$

43. $1 - \cos A = \dfrac{\cos(b - c) - \cos a}{\sin b \sin c}$

From the identity $\sin^2 \theta = (1 - \cos 2\theta)/2$, we can find that $1 - \cos A = 2\sin^2(A/2)$:

$$2\sin^2(A/2) = \frac{\cos(b - c) - \cos a}{\sin b \sin c}$$

From the identity $\cos \theta - \cos \phi = -2 \sin[(\theta + \phi)/2] \sin[(\theta - \phi)/2]$ we can rewrite the numerator of the right hand side:

$$2\sin^2(A/2) = \frac{-2\sin[(a + b - c)/2] \times \sin[(-a + b - c)/2]}{\sin b \sin c}$$

$$\sin^2(A/2) = \frac{\sin[(a + b - c)/2] \sin[(a - b + c)/2]}{\sin b \sin c}$$

Since $s = (a + b + c)/2$, this can be rewritten:

$$\sin(A/2) = \sqrt{\frac{\sin(s - b) \sin(s - c)}{\sin b \sin c}}$$

From symmetry, these formulas also are true:

$$\sin(B/2) = \sqrt{\frac{\sin(s - a) \sin(s - c)}{\sin a \sin c}}$$

$$\sin(C/2) = \sqrt{\frac{\sin(s - a) \sin(s - b)}{\sin a \sin b}}$$

44. $\sin(A/2) \cos(B/2)$

$$= \sqrt{\frac{\sin(s - b) \sin(s - c)}{\sin b \sin c} \frac{\sin s \sin(s - b)}{\sin c \sin a}}$$

$$= \frac{\sin(s - b)}{\sin c} \sqrt{\frac{\sin s \sin(s - c)}{\sin a \sin b}}$$

$$= \frac{\sin(s - b) \cos(C/2)}{\sin c}$$

Following the same procedure we can find:

$$\sin(B/2) \cos(A/2) = \frac{\sin(s - a) \cos(C/2)}{\sin c}$$

Then:

$$\sin(A/2) \cos(B/2) - \sin(B/2) \cos(A/2)$$

$$= \frac{[\sin(s - b) - \sin(s - a)] \cos(C/2)}{\sin c}$$

The right hand side can be rewritten using the difference formula for sines:

$$\frac{[\sin(s-b)-\sin(s-a)]\cos(C/2)}{\sin c}$$

$$=\frac{2\cos[(2s-a-b)/2]\sin[(a-b)/2]\cos(C/2)}{\sin c}$$

Since $2s - a - b = c$, and $\sin c = 2\sin(c/2)\cos(c/2)$, we have:

$$=\frac{2\cos(c/2)\sin[(a-b)/2]\cos(C/2)}{2\sin(c/2)\cos(c/2)}$$

$$=\frac{\sin[(a-b)/2]\cos(C/2)}{\sin(c/2)}$$

The left hand side can be rewritten:

$$\sin(A/2)\cos(B/2) - \sin(B/2)\cos(A/2) = \sin[(A-B)/2]$$

Therefore:

$$\frac{\sin[(A-B)/2]}{\cos(C/2)} = \frac{\sin[(a-b)/2]}{\cos(c/2)}$$

There are other formulas that can be derived in a similar manner:

$$\frac{\cos[(A-B)/2]}{\sin(C/2)} = \frac{\sin[(a+b)/2]}{\sin(c/2)}$$

$$\frac{\sin[(A+B)/2]}{\cos(C/2)} = \frac{\cos[(a-b)/2]}{\cos(c/2)}$$

$$\frac{\cos[(A+B)/2]}{\sin(C/2)} = \frac{\cos[(a+b)/2]}{\cos(c/2)}$$

45. $$\tan[(A-B)/2] = \frac{\sin[(A-B)/2]}{\cos[(A-b)/2]}$$

$$=\frac{\dfrac{\sin[(a-b)/2]\cos(C/2)}{\sin(c/2)}}{\dfrac{\sin[(a+b)/2]\sin(C/2)}{\sin(c/2)}}$$

$$=\frac{\sin[(a-b)/2]\operatorname{ctn}(C/2)}{\sin[(a+b)/2]}$$

$$\frac{\tan[(A-B)/2]}{\cos(C/2)} = \frac{\sin[(a-b)/2]}{\sin[(a+b)/2]}$$

Chapter 15

1. For small x, $\sin x = x$

2. For small x, $\cos x = 1 + x^2/2$

3.

Number of terms	x 0.1	1	π/2	π/6	π/4
2	0.9983	0.8333	0.9248	0.4997	0.7047
3	0.9983	0.8417	1.0045	0.5000	0.7071
4	0.9983	0.8415	0.9998	0.5000	0.7071
5	0.9983	0.8415	1.0000	0.5000	0.7071
sin x	0.9983	0.8415	1.0000	0.5000	0.7071

As you can see, for small values of x this series converges very rapidly to the true value for $\sin x$.

4. Since $\tan(\pi/4) = 1$, it follows that $\pi/4 = \arctan 1$. Therefore,

$$\frac{\pi}{4} = 1 - \frac{1}{3} + \frac{1}{5} - \frac{1}{7} + \frac{1}{9} - \cdots$$

Glossary

abscissa Abscissa means x coordinate.

absolute value The absolute value of a real number a is

$$|a| = a \quad \text{if } a > 0; \qquad |a| = -a \quad \text{if } a < 0$$

The absolute value of a complex number $a + bi$ is $\sqrt{a^2 + b^2}$.

acute angle An acute angle is an angle that measures less than $90°$.

acute triangle An acute triangle is a triangle that contains three acute angles.

altitude An altitude of a triangle is a line segment connecting one vertex of the triangle to the line containing the opposite side; it is perpendicular to the opposite side.

amplitude The amplitude of the periodic function $A \sin t$ is A.

angle An angle is the union of two rays with a common end point.

arc An arc is a curve that is part of a circle.

arccos, arccsc, arcctn, arcsec, arcsin, arctan These are the inverse functions for the six trigonometric functions.

argument The argument of a function is the independent variable that is put into the function.

asymptote An asymptote is a straight line that is a close approximation to a particular curve as the curve goes off to infinity in one direction. The curve comes very close to the asymptote line, but it never touches it.

axis (1) The x axis in cartesian coordinates is the line $y = 0$. The y axis is the line $x = 0$. (2) The axis of a figure is a line about which the figure is symmetrical.

binomial A binomial is the sum of two terms, such as $2a^2 + 10xy$.

Cartesian coordinates A Cartesian coordinate system is a system in which each point on a plane is identified by an ordered pair of numbers representing its distances from two perpendicular lines, or in which each point in space is similarly identified by its distances from three perpendicular planes.

central angle A central angle is an angle whose vertex is at the center of a circle.

circle A circle is a set of points in a plane that are all the same distance from a given point.

circumference The circumference of a closed curve (such as a circle) is the total distance around the outer edge of the curve.

coefficient Coefficient is a technical term for something that multiplies something else, usually a fixed number multiplying a variable.

cofunction The cofunction of the sine function is the cosine function; the cofunction of the tangent function is the cotangent function; and the cofunction of the secant function is the cosecant function.

common logarithm A common logarithm is a logarithm to the base 10.

complex number A complex number is formed by adding a pure imaginary number to a real number. The general form of a complex number is $a + bi$, where a and b are real numbers and $i = \sqrt{-1}$.

congruent Two triangles are congruent if they have the same shape and size.

conic sections The four curves—circle, ellipse, parabola, and hyperbola—are called conic sections because they can be formed by the intersection of a plane with a right circular cone.

conjugate The conjugate of a complex number is formed by reversing the sign of the imaginary part.

coordinates The coordinates of a point are a set of numbers that identify the location of that point.

cos This is the abbreviation for cosine.

cosecant The cosecant function is the reciprocal of the sine function.

cosine Cosine is a trigonometric function. For an acute angle in a right triangle, the cosine is the length of the adjacent side divided by the length of the hypotenuse. For the general definition, see Chapter 5.

cotangent Cotangent is a trigonometric function abbreviated ctn:

$$\text{ctn} \, x = \tan\left(\frac{\pi}{2} - x\right) = \frac{1}{\tan x}$$

degree A degree is a unit of measure for angles. One degree is equal to $\frac{1}{360}$th of a full rotation. The symbol for degree is a little raised circle: $°$.

denominator The denominator is the bottom part of a fraction.

dependent variable The dependent variable stands for any of the set of output numbers of a function. In the equation $y = f(x)$, y is the dependent variable and x is the independent variable.

diameter The diameter of a circle is the length of a line segment joining two points on the circle and passing through the center.

discriminant See quadratic formula.

eccentricity The eccentricity of a conic section is a number that indicates the shape of the conic section.

equation An equation is a statement that says two mathematical expressions have the same value.

equilateral triangle An equilateral triangle is a triangle with three equal sides. An equilateral triangle has three 60° angles.

even number An even number is a number that is divisible evenly by 2, such as 2, 4, 6, 8, 10, ...

exponent An exponent is a number that indicates the operation of repeated multiplication.

factor A factor is one of two or more expressions that can be multiplied together to get a given expression; they are said to be factors of that expression.

factorial The factorial of a positive integer is the product of all the integers from 1 up to that number. The exclamation point! is used to designate the factorial. For example, $4! = 4 \times 3 \times 2 \times 1 = 24$.

frequency The frequency of a wave is the number of crests that pass a given point each second.

function A function is a rule that turns one number into another number.

geometric series A geometric series is a sum of terms of the form $a + ar + ar^2 + ar^3 + \ldots + ar^{n-1}$.

graph The graph of an equation is the set of points (with values given by coordinates) that make the equation true.

hyperbola A hyperbola is the set of points in a plane such that the difference between the distances to two fixed points is a constant.

hypotenuse The hypotenuse is the side of a right triangle that is opposite the right angle.

i i is the basic unit for imaginary numbers. i is defined by the equation $i^2 = -1$.

identity An identity is an equation that is true for all possible values of the unknowns it contains.

imaginary number An imaginary number (or a pure imaginary number) is a number of the form bi, where b is a real number and $i = \sqrt{-1}$.

independent variable The independent variable stands for any of the set of input numbers to a function.

integers The set of integers contains zero, the natural numbers, and the negatives of the natural numbers:

. . . , $-6, -5, -4, -3, -2, -1, 0, 1, 2, 3, 4, 5, 6, \ldots$

inverse function The inverse function of a function is the function that does exactly the opposite of the original function. See Chapter 10 for a description of the inverse trigonometric functions.

irrational number An irrational number is a real number that cannot be expressed as the ratio of two integers.

isosceles triangle An isosceles triangle has two equal sides.

logarithm The equation $y = a^x$ can be written $x = \log_a y$, which means "x is the logarithm of y to the base a."

major axis The major axis of an ellipse is the line segment joining two points on the ellipse that passes through the two focus points.

minute A minute is a unit of measure for angles. One minute $= \frac{1}{60}$ degree.

numerator The numerator is the top part of a fraction.

obtuse angle An obtuse angle is an angle larger than a 90° angle.

obtuse triangle An obtuse triangle is a triangle that contains one obtuse angle.

odd number An odd number is a natural number that is not divisible by 2, such as 1, 3, 5, 7, 9, ...

ordinate The ordinate of a point is another name for y coordinate.

origin The origin is the point (0, 0) of a Cartesian coordinate system.

parabola A parabola is the set of all points that are equally distant from a fixed point (called the focus) and a fixed line (called the directrix).

periodic A periodic function is a function that keeps repeating the same pattern of values. Formally, a function $f(x)$ is periodic if there exists a number p such that $f(x + p) = f(x)$ for all x.

perpendicular Two lines are perpendicular if they meet so as to form a right angle.

pi The Greek letter π (pi) is used to represent the ratio between the circumference of a circle and its diameter:

$$\pi = \frac{\text{circumference}}{\text{diameter}}$$

This ratio is the same for any circle. π is an irrational number with the decimal approximation 3.14159.

plane A plane is a flat surface (like a table top) that stretches to infinity in all directions.

polar coordinates Any point in a plane can be identified by listing its distance from a specified origin and the angle between a specified 0° direction and the line connecting the point to the origin. This system is called the polar coordinate system. See Chapter 11.

polynomial A polynomial in x is an algebraic expression of the form $a_n x^n + a_{n-1} x^{n-1} + \ldots + a_2 x^2 + a_1 x + a_0$ where $a_n, \ldots, a_0$ are constants that are the coefficients of the polynomial. The degree of the polynomial is the highest power of the variable that appears (in this case n).

power A power of a number indicates repeated multiplication. For example, the third power of 5 is $5^3 = 5 \times 5 \times 5 = 125$.

Pythagorean theorem The Pythagorean theorem relates the three sides of a right triangle: $c^2 = a^2 + b^2$, where c is the side opposite the right angle (called the hypotenuse) and a and b are the sides adjacent to the right angle.

quadrant The x and y axes in a Cartesian coordinate system divide a plane into four quadrants. The quadrant where x and y are both positive is called the first quadrant; where x is negative and y is positive is called the second quadrant; where x and y are both negative is called the third quadrant; and where x is positive and y is negative is called the fourth quadrant.

quadratic equation A quadratic equation in one variable, x, is an equation of the form

$$ax^2 + bx + c = 0$$

A quadratic equation in two variables, x and y, is an equation of the form

$$Ax^2 + Bxy + Cy^2 + Dx + Ey + F = 0$$

The quantity $B^2 - 4AC$ is called the discriminant because its value determines the nature of the solution. See Chapter 13.

quadratic formula The quadratic formula states that the solutions to the equation $ax^2 + bx + c = 0$ are

$$\frac{-b \pm \sqrt{b^2 - 4ac}}{2a}$$

The quantity $b^2 - 4ac$ is called the discriminant.

radian measure Radian measure is a system for measuring angles in which one complete rotation measures 2π radians.

radical The radical symbol $\sqrt{}$ is used to indicate a root of a number.

radius The radius of a circle is the distance from the center of the circle to a point on the circle.

rational number A rational number is any number that can be expressed as the ratio of two integers.

ray A ray is like half of a line: it has one end point, and then goes off forever in a straight line.

real numbers The set of real numbers is the set of all numbers that can be represented by a point on a number line. The set of real numbers includes all rational numbers and all irrational numbers.

reciprocal The reciprocal of a number a is $1/a$.

right angle A right angle is an angle that measures 90°.

right triangle A right triangle is a triangle that contains one right angle.

root The process of taking a root of a number is the opposite of raising that number to a power.

scalene A scalene triangle is a triangle in which no two sides have equal length.

scientific notation Scientific notation is a short way of writing very large or very small numbers. A number in scientific notation is expressed as a number between 1 and 10 multiplied by a power of 10.

secant The secant function is the reciprocal of the cosine function.

second A second is a unit of measure for angles. One second = $\frac{1}{60}$ minute = $\frac{1}{3600}$ degree.

semimajor axis The semimajor axis of an ellipse is equal to one-half the longest distance across the ellipse.

semiminor axis The semiminor axis of an ellipse is equal to one-half the shortest distance across the ellipse.

simultaneous equations A system of simultaneous equations is a group of equations that must all be true at the same time.

sin Abbreviation for sine.

sine A trigonometric function. In a right triangle the sine of an acute angle is equal to the length of the opposite side divided by the length of the hypotenuse. For the general definition of the sine function see Chapter 5.

slope The slope of a line is a number that measures how steep the line is. A horizontal line has a slope of zero. A vertical line has an infinite slope.

solution The solution of an equation is the value(s) of the variable(s) contained in that equation that make(s) the equation true.

square The square of a number is found by multiplying that number by itself.

square root The square root of a number a (written $\sqrt{a}$) is a number that, when multiplied by itself, gives a.

substitution property The substitution property states that, if $a = b$, then you can replace the expression a anywhere it appears by the expression b if you want to.

sum The sum is the result when two or more numbers are added.

tan This is the abbreviation for tangent.

tangent Tangent is a trigonometric function. The tangent of an acute angle in a right triangle is equal to the length of the opposite side divided by the length of the adjacent side. For the general definition of the tangent function see Chapter 5.

triangle A triangle consists of three line segments joined end to end.

wavelength The wavelength of a wave is the distance between crests.

Appendix

In practice, we can never come up with exact answers for trigonometry problems, for two reasons. First, we cannot work with exact values of trigonometric functions. Except for a few special cases (such as $\sin 30° = \frac{1}{2}$), it would require an infinite number of decimal digits to express the value of a trigonometric function exactly. Therefore, we use approximations containing a finite number of decimal places.

Second, no measurement is perfectly accurate. If you measure the length of a line segment or the size of an angle, there will always be some uncertainty in your measurement. If you are more careful and have better equipment, you can come up with a more precise measurement, but it will never be perfect. Since any numerical answer we give for a trigonometry problem is an approximation, the question is: How many digits should we display in the result? That depends on the number of *significant digits*.

Suppose you quickly estimate the distance to a building and give the result as 500 feet. Often this does not mean that the result is exactly 500 feet. The zeros at the end of the number are said to be nonsignificant, and so your estimated measurement has only one significant digit (the 5 at the front). This implies that you think the true value is really somewhere between 450 feet and 550 feet. If the true value is 430 feet, then you would be misleading someone if you gave the result as 500 feet.

Now suppose you conduct a more careful measurement and report the distance as 530 feet. As before, the zero at the end is nonsignificant, and so your measurement has two significant digits. You are implying that the true value is between 525 feet and 535 feet.

After further careful measurement, you give the result as 534 feet. Now your measurement has three significant digits, and you are implying that the true value is between 533.5 and 534.5.

More careful measurement allows you to give the value as 534.6. Now your measurement has four significant digits, and you are implying that the true value is between 534.55 and 534.65.

If you report the measurement as 534.60, you will have five significant digits. Note that any zero that occurs after a decimal point counts as a significant digit.

In the number 230,405,200 there are seven significant digits. The two zeros at the end of the number are not significant, but the zeros that occur between the other digits are significant.

In the number 500.00, there are five significant digits. If there are any digits to the right of the decimal point, then all the zeros in the number are significant digits.

In the middle of a calculation, you should keep as many decimal places as your calculator stores. If you round off the result before you come to the final answer, your result will be less accurate. For example, suppose you need to find $50.00 \times \tan 40.00°$. Enter 40 into your calculator and press the tan button. The result might have 12 decimal places, such as 0.839099631177. Multiply this value by 50 to give 41.9549815588. You should never include 12 decimal places in the final answer you display because that implies that your measurements have been superhumanly accurate. Round this result to four significant digits: 41.95. In this book we usually assume that measurements are accurate enough to allow three or four significant digits in the results of calculations. However, we usually write whole numbers without the zeros trailing the decimal point (that is, we write 50 instead of 50.00) to avoid clutter.

Computer spreadsheets, such as Lotus 1-2-3, Microsoft Excel, and Microsoft Works, are very valuable computational tools that can help with trigonometry problems. Although the exact details of spreadsheets vary, the basic idea is that a spreadsheet allows you to enter formulas that depend on other values. If you change these values, the result of the formula will automatically be recalculated.

A simple spreadsheet: finding height given angle of elevation and distance

In Chapter 3 we calculated the height of an object when we know its distance and its angle of elevation. We can set up a simple spreadsheet.

The numbers at the far left are the spreadsheet row numbers, and the letters at the top are the column labels. In this example, we are calculating the height of an object at a distance of 100 with an angle of elevation of 20 degrees. (The unit you use to measure the distance doesn't matter; the result for the height will be in the same unit you used for the distance.) Enter the formula as shown in cell B3. The angle in B1 is given in degrees, but it is converted to radians by multiplying by π and dividing by 180. When you look at cell B3 in your spreadsheet, you see the number that is the result of the formula, not the formula itself. If you change the numbers in cells B1 and B2, the result for the height will be recalculated automatically.

	A	B
1	Angle of elevation:	20
2	Distance:	100
3	Height: =B2*TAN(B1*3.14159/180)	

Finding the distance between cities

The next spreadsheet example is a bit more complicated: it contains formulas that find the distance along a great circle course between two cities (see Chapter 14). The latitudes are entered in cells C2 and D2. Next, the computer calculates the complement of the latitude for each city (labeled s) and then converts this value to radians (cells C3 and D3). Cell C7 contains the formula for D, the difference between the longitudes converted to radians. Cell C11 contains the formula that finds the distance between the cities (see page 238).

The formula in cell C11 is:
`=C9*ACOS(COS(C3)*COS(D3)+SIN(C3)*SIN(D3)*COS(C7))`

	B	C	D
1		City 1	City 2
2	latitude(deg)	41	34
3	s(rad)	=(90-C2)*3.14159/180	=(90-D2)*3.14159/180
4	longitude(deg)	286	242
5			
6	D (deg)	=D4-C4	
7	D(rad)	=C6*3.14159/180	
8			
9	r (km)	6375	
10			
11	distance(km)	*	
12	distance(miles)	=C11*0.621371	

ACOS is the abbreviation for the arccosine
function. The example given shows New York (latitude
41, longitude 286 east) and Los Angeles (latitude 34,
longitude 242 east). To find the distance between
another pair of cities, all you need to do is change the
numbers in cells C2, D2, C4, and D4.

Solving triangles

Here is a spreadsheet that uses the laws of sines
and cosines to solve for the remaining parts of a
triangle if you know three of the parts. (See Chapter 7,
page 109). We need to consider five different situations.
In each case we follow our traditional notation and use
a, b, and c to represent the sides and A, B, and C to
represent the angles. (Don't confuse these letters with
the A, B, C, . . . that are the labels of the spreadsheet
columns. The column label letters are shown at the top
of each table below.) In each table, a 0 represents a cell
where you have to fill in the appropriate value. A *
indicates a cell that contains a formula; these formulas
are displayed below the table. Cell G2 contains the
formula =180/3.14159; it is used to convert between
radians and degrees when necessary.

Two sides and angle between them are known:

	A	B	C	D	E	F
2	a	b	c	A	B	C
3	0	0	*	*	*	0

Formula in cell C3:
`=SQRT(A3^2+B3^2-2*A3*B3*COS(F3/G2))`

Formula in cell D3:
`=G2*ACOS((B3^2+C3^2-A3^2)/(2*B3*C3))`

Formula in cell E3:
`=G2*ACOS((A3^2+C3^2-B3^2)/(2*A3*C3))`

One side and the two angles next to that side are known:

	A	B	C	D	E	F
6	a	b	c	A	B	C
7	0	*	*	*	0	0

Formula in cell B7:
`=A7*SIN(E7/G2)/SIN(D7/G2)`

Formula in cell C7:
`=A7*SIN(F7/G2)/SIN(D7/G2)`

Formula in cell D7:
`=180-(E7+F7)`

Three sides are known:

	A	B	C	D	E	F
10	a	b	c	A	B	C
11	0	0	0	*	*	*

Formula in cell D11:
`=G2*ACOS((B11^2+C11^2-A11^2)/(2*B11*C11))`

Formula in cell E11:
`=G2*ACOS((A11^2+C11^2-B11^2)/(2*A11*C11))`

Formula in cell F11:
`=G2*ACOS((A11^2+B11^2-C11^2)/(2*A11*B11))`

Two sides and one angle (not between the two sides) are known; triangle is known to be acute

	A	B	C	D	E	F
15	a	b	c	A	B	C
16	*	0	0	*	0	*

Formula in cell A16:
`=SQRT(B16^2+C16^2-2*B16*C16*COS(D16/G2))`

Formula in cell D16:
`=180-(E16+F16)`

Formula in cell F16:
`=G2*ASIN(C16*SIN(E16/G2)/B16)`

Two sides and one angle (not between the two sides) are known; triangle is known to be obtuse

	A	B	C	D	E	F
20	a	b	c	A	B	C
21	*	0	0	*	0	*

Formula in cell A21:
`=SQRT(B21^2+C21^2-2*B21*C21*COS(D21/G2))`

Formula in cell D21:
`=180-(E21+F21)`

Formula in cell F21:
`=180-G2*ASIN(C21*SIN(E21/G2)/B21)`

Once you have the spreadsheet set up, it is a good idea to lock, or protect, all the cells with formulas so that you cannot accidentally change them while you are filling in the numbers in other cells.

You might also see if your spreadsheet allows you to assign a name to a cell. Then it would be possible for you to write your formulas in a more meaningful form. In the formulas above, all quantities are identified by their cell addresses rather than by a meaningful name.

Graphing with a spreadsheet

A computer spreadsheet can easily be used to make a graph of a curve such as $y = \sin x$. Set up the first column as follows:

	A	B
1	x	y=sin x
2	0	=sin(a2)
3	0.1	″
4	0.2	″
5	0.3	″
6	0.4	″
	⋮	
322	3.2	″

To save space, not all the rows of the spreadsheet are shown on this page. You don't have to type in all these numbers; your spreadsheet has a way to fill in the values automatically. In Microsoft Works or Excel, enter the first two values (0 and 0.1), highlight both of them, and then drag the lower right corner of the highlighted region down to fill in the rest of the values.

Next, enter the label

```
y=sinx
```

in cell B1 and then enter the formula

```
=sin(a2)
```

in cell B2. Copy this formula down to the other cells in column B. (Each cell with a ″ means that its formula is found by copying down from the formula above it.) Next, use the spreadsheet's graphing command to create an *xy* (or scatter) graph, with column A representing the *x* values and column B representing the *y* values. The computer automatically plots all these points to create the graph. Try experimenting by changing the range of values graphed to see how the graph changes.

Creating a polar coordinate graph with spreadsheets

Enter the values 0 to 6.2 for θ into column A as shown:

	A	B	C	D
1	theta	r	x	y
2	0	=1−cos(a2)	=b2*cos(a2)	=b2*sin(a2)
3	0.1	″	″	″
4	0.2	″	″	″
5	0.3	″	″	″
6	0.4	″	″	″
⋮				
622	6.2	″	″	″

In cell B2, enter the formula you wish to graph. For example, enter

```
=1−cos(a2)
```

to create the heart-shaped curve in Figure 11-10. Note that each row in this spreadsheet represents a point on the graph. Columns A and B give the polar coordinates, and columns C and D represent the rectangular coordinates of that point. Enter the formula

```
=b2*cos(a2)
```

into cell C2 and enter the formula

```
=b2*sin(a2)
```

into cell D2. These formulas will convert the polar coordinates to rectangular coordinates. Copy the formulas in cells B2, C2, and D2 into the remaining cells in those columns. Now, create an *xy* (scatter) graph using the *x* and *y* coordinates in columns C and

D. Whenever you want to make a graph of a different polar coordinate curve, all you need to do is change the formula in cell B2 and copy that formula down to the remaining cells in column B.

Graphing an ellipse in a rotated coordinate system

We can make a graph of the rotated translated ellipse shown in Figure 13-13. (page 210). The spreadsheet will be set up so that you can also use it to graph any ellipse when you are given an equation of the form:

$$Ax^2 + Bxy + Cy^2 + Dx + Ey + F = 0$$

With a little rearrangement you could also make the spreadsheet graph parabolas or hyperbolas. The values of the original coefficients A to F are entered into cells C2 to C7. The angle of rotation is calculated in cell C10, followed by calculation of the coefficients in the rotated coordinate system. Then the translation parameters h and k are calculated.

To create the graph, column B contains the values of x'' (labeled x2 because it is awkward to use a prime symbol as part of a label in a computer spreadsheet), which run from the smallest possible value of x2 (called –x2max) up to the largest possible value (x2max) and then back down to –x2max. The positive value of y2 is calculated in cells C32 to C131; the negative value of y2 is in cells C132 to C230.

Columns D and E contain the values of x1 and y1; these are the values obtained from the translation formula x1=x2-h and y1=y2-k. (These equations would be written $x' = x'' - h$ and $y' = y'' - k$ in the notation in Chapter 13.) Columns F and G contain the values of x and y after applying the coordinate rotation. To create the graph, set up an xy (scatter) graph for the values in these last two columns.

A dollar sign $ in a cell address means that the address will not be changed when the formula is copied to another cell. See the manual for your spreadsheet for more information about this feature.

Here is the spreadsheet:

	B	C	D
1		original coefficients:	
2	A	0.0474	
3	B	−0.02114	
4	C	0.0551	
5	D	−0.15344	

	B	C	D
6	E	−1.17562	
7	F	6.0625	
8			
9		angle of rotation:	
10	Theta:	=ATAN(C3/(C2-C4))/2	
11	in degrees:	=C10*180/3.14159	
12	cos(theta)	=COS(C10)	
13	sin(theta)	=SIN(C10)	
14			
15		coefficients in rotated coordinate system:	
16	A1	=C2*C12^2+C4*C13^2+C3*C12*C13	
17	C1	=C2*C13^2+C4*C12^2-C3*C12*C13	
18	D1	=C5*C12+C6*C13	
19	E1	=−C5*C13+C6*C12	
20	F1	=C7	
21			
22		translation parameters	
23	H	=C18/(2*C16)	
24	K	=C19/(2*C17)	
25	F2	=C20−(C18^2/(4*C16)+C19^2/(4*C17))	
26			
27	x2max	=SQRT(−C25/C16)	

	B	C	D	E	F	G
31	x2	y2	x1	y1	x	y
32	=−C27	*	*	*	=D32*$C$12−E32*$C$13	=E32*$C$12+D32*$C$13
33	*	"	"	"	"	"
34	"	"	"	"	"	"
			⋮			
131	*	*	"	"	"	"
132	*	*	"	"	"	"
			⋮			
230	"	"	"	"	"	"

Formula in cell B33:
=B32+2*$C$27/100

Formula in cell B131:
=B130+2*$C$27/100

Formula in cell B132:
=B131−2*$C$27/100

Formula in cell C32:
=SQRT((−$C$25−$C$16*B32^2)/$C$17)

Formula in cell C131:
=SQRT((−$C$25−$C$16*B131^2)/$C$17)

Formula in cell C132:
=−SQRT((−$C$25−$C$16*B132^2)/$C$17)

Formula in cell D32:
=B32−$C$23

Formula in cell E32:
=C32−$C$24

Note how the plus sign changes to a minus sign between cells B131 and B132; this is when the values x2 in column B stop increasing and start decreasing. Also, note the sign change in the values for y2 between cells C131 and C132.

Summary of Trigonometric Formulas

Trigonometric functions for right triangles

Let A be one of the acute angles in a right triangle. Then,

$$\sin A = \frac{\text{opposite side}}{\text{hypotenuse}}$$

$$\cos A = \frac{\text{adjacent side}}{\text{hypotenuse}}$$

$$\tan A = \frac{\text{opposite side}}{\text{adjacent side}}$$

Trigonometric functions: General definition

Consider a point (x, y) in a Cartesian coordinate system. Let r be the distance from that point to the origin, and let A be the angle between the x axis and the line connecting the origin to that point. Then,

$$\sin A = \frac{y}{r}$$

$$\cos A = \frac{x}{r}$$

$$\tan A = \frac{y}{x}$$

Radian measure

$$\pi \text{ rad} = 180°$$

Special values

Degrees	Radians	sin	cos	tan
0°	0	0	1	0
30°	$\frac{\pi}{6}$	$\frac{1}{2}$	$\frac{\sqrt{3}}{2}$	$\frac{1}{\sqrt{3}}$
45°	$\frac{\pi}{4}$	$\frac{1}{\sqrt{2}}$	$\frac{1}{\sqrt{2}}$	1
60°	$\frac{\pi}{3}$	$\frac{\sqrt{3}}{2}$	$\frac{1}{2}$	$\sqrt{3}$
90°	$\frac{\pi}{2}$	1	0	Undefined (infinite)

Trigonometric identities

These equations are true for every allowable value of A and B.

Reciprocal functions

$$\sin A = \frac{1}{\csc A} \quad \csc A = \frac{1}{\sin A}$$

$$\cos A = \frac{1}{\sec A} \quad \sec A = \frac{1}{\cos A}$$

$$\tan A = \frac{1}{\operatorname{ctn} A} \quad \operatorname{ctn} A = \frac{1}{\tan A}$$

Cofunctions (radian form)

$$\sin A = \cos\left(\frac{\pi}{2} - A\right) \quad \cos A = \sin\left(\frac{\pi}{2} - A\right)$$

$$\tan A = \operatorname{ctn}\left(\frac{\pi}{2} - A\right) \quad \operatorname{ctn} A = \tan\left(\frac{\pi}{2} - A\right)$$

$$\sec A = \csc\left(\frac{\pi}{2} - A\right) \quad \csc A = \sec\left(\frac{\pi}{2} - A\right)$$

Negative angle relations

$$\sin(-A) = -\sin A$$
$$\cos(-A) = -\cos A$$
$$\tan(-A) = -\tan A$$

Quotient relations

$$\tan A = \frac{\sin A}{\cos A}$$

$$\operatorname{ctn} A = \frac{\cos A}{\sin A}$$

Supplementary angle relations

The angles A and B are supplementary angles if $A + B = \pi$.

$$\sin(\pi - A) = \sin A$$
$$\cos(\pi - A) = -\cos A$$
$$\tan(\pi - A) = -\tan A$$

Pythagorean identities

$$\sin^2 A + \cos^2 A = 1$$
$$\tan^2 A + 1 = \sec^2 A$$
$$\operatorname{ctn}^2 A + 1 = \csc^2 A$$

Functions of the sum of two angles

$$\sin(A + B) = \sin A \cos B + \sin B \cos A$$
$$\cos(A + B) = \cos A \cos B - \sin A \sin B$$

$$\tan(A + B) = \frac{\tan A + \tan B}{1 - \tan A \tan B}$$

Functions of the difference of two angles

$$\sin(A - B) = \sin A \cos B - \sin B \cos A$$

$$\cos(A - B) = \cos A \cos B + \sin A \sin B$$

$$\tan(A - B) = \frac{\tan A - \tan B}{1 + \tan A \tan B}$$

Double-angle formulas

$$\sin(2A) = 2 \sin A \cos A$$

$$\cos(2A) = \cos^2 A - \sin^2 A$$

$$= 1 - 2 \sin^2 A$$

$$= 2 \cos^2 A - 1$$

$$\tan(2A) = \frac{2 \tan A}{1 - \tan^2 A}$$

Squared formulas

$$\sin^2 A = \frac{1}{2}(1 - \cos 2A)$$

$$\cos^2 A = \frac{1}{2}(1 + \cos 2A)$$

Half-angle formulas

$$\sin \frac{A}{2} = \pm \sqrt{\frac{1 - \cos A}{2}}$$

$$\cos \frac{A}{2} = \pm \sqrt{\frac{1 + \cos A}{2}}$$

$$\tan \frac{A}{2} = \pm \sqrt{\frac{1 - \cos A}{1 + \cos A}}$$

Product formulas

$$\sin A \cos B = \frac{1}{2}[\sin(A + B) + \sin(A - B)]$$

$$\cos A \sin B = \frac{1}{2}[\sin(A + B) - \sin(A - B)]$$

$$\cos A \cos B = \frac{1}{2}[\cos(A + B) - \cos(A - B)]$$

$$\sin A \sin B = \frac{1}{2}[\cos(A + B) - \cos(A - B)]$$

Sum formulas

$$\sin A + \sin B = 2 \sin \frac{A + B}{2} \cos \frac{A - B}{2}$$

$$\cos A + \cos B = 2 \cos \frac{A + B}{2} \cos \frac{A - B}{2}$$

Difference formulas

$$\sin A - \sin B = 2\cos\frac{A+B}{2}\sin\frac{A-B}{2}$$

$$\cos A - \cos B = -2\sin\frac{A+B}{2}\sin\frac{A-B}{2}$$

Formulas for triangles

Let a be the side of a triangle opposite angle A, let b be the side opposite angle B, and let c be the side opposite angle C.

Law of cosines

$$c^2 = a^2 + b^2 - 2ab\cos C$$

Law of sines

$$\frac{a}{\sin A} = \frac{b}{\sin B} = \frac{c}{\sin C}$$

Index